A Primer
ON Theory AND Operation
OF Linear Accelerators
IN Radiation Therapy

SECOND EDITION

C.J. KARZMARK, PH.D.
PROFESSOR EMERITUS, RADIATION ONCOLOGY
DEPARTMENT OF RADIATION ONCOLOGY
STANFORD UNIVERSITY SCHOOL OF MEDICINE
STANFORD, CALIFORNIA

&

ROBERT J. MORTON, M.S.
PRESIDENT
QUALITY AND REGULATORY SERVICES, INC.
WALNUT CREEK, CALIFORNIA

Medical Physics Publishing
Madison, Wisconsin

Originally published by the Bureau of Radiological Health, 1981.
Published as *A Primer on Theory and Operation of Linear Accelerators in Radiation Therapy,* first edition, in 1989 by Medical Physics Publishing.

Library of Congress Cataloging-in-Publication Data

Karzmark, C.J.
    A primer on theory and operation of linear accelerators in radiation therapy / by C. J. Karzmark and Robert J. Morton. — 2nd ed.
      p.   cm.
    Includes bibliographical references and index.
    ISBN 0-944838-66-9 (paper)
    1. Radiotherapy—Instruments. 2. Linear accelerators. 3. Medical physics. I. Morton, Robert J. II. Title.
    [DNLM: 1. Particle Accelerators. 2. Radiotherapy—instrumentation. QZ 269 K18p 1996]
    RM862.5.K37   1996
    615.8′42—dc20
    DNLM/DLC
    for Library of Congress   96-24382
                                          CIP

Design and Composition by Colophon Typesetting

Medical Physics Publishing
4513 Vernon Blvd.
Madison, WI 53705
608-262-4021

Contents

Preface

Electron linear accelerators evolved from the microwave radar developments of World War II. The klystron tube, invented at Stanford University, provided a vital source of microwave power for radar then as it does now. In the late 1940s, the high-power klystron and the microwave principles incorporated into its design were used to construct and power an electron linear accelerator, or linac, for use in physics research and later for industrial radiography. By the mid-1950s, a linac suitable for treating deep-seated tumors was built in the Stanford Microwave Laboratory and installed at Stanford Hospital, which was located in San Francisco at that time. It served as a prototype for commercial units that were built later.

Since that time medical linear accelerators have gained in popularity as major radiation therapy devices, but few basic training materials on their operation have been produced for use by medical professionals. Dr. C. J. Karzmark, a radiological physicist at Stanford University, has been involved with medical linacs since their development, and he agreed to collaborate with Robert Morton of the Center for Devices and Radiological Health (formerly the Bureau of Radiological Health), U.S. Food and Drug Administration, in writing this primer on the operation of medical linear accelerators. The primer was originally published by the U.S. Department of Health and Human Services in December 1981, as FDA 82-8181. This publication provides an overview of the components of the linear accelerator and how they function and interrelate. The auxiliary systems necessary to maintain the operation of the linear accelerator are also described. The primer will promote an understanding of the safe and effective use of these devices. It was produced in cooperation with the Division of Resources, Centers, and Community Activities of the National Cancer Institute, and is intended for students of radiation therapy, radiological physics, radiation oncology, and radiation control.

For ease in understanding, much of the text describes the components as they appear in a specific electron linear accelerator treatment unit, Varian Associates' Clinac 18. This choice in no way constitutes an endorsement of this particular equipment. Variations in design do occur and several are described in Appendix A. A videotape, titled "The Theory and Operation of the Linear Accelerator in Radiation Therapy," has been produced in conjunction with this primer and can be ordered from Medical Physics Publishing.

The current revision takes cognizance of significant advances occurring in radiotherapy linacs since the original publication. Again, the level of treating these advances is simplified so that the audience of radiation therapists, as well as physicians, engineers, and physicists, can benefit. A new Section 9, "Dual X-Ray Energy Mode Accelerators," describes these versatile new units that provide two x-ray and several electron beams for a variety of clinical situations. Providing these various treatment modalities requires changes in how the standing wave and traveling wave accelerator structures are energized with microwave power. Section 10, "Bending Magnet," has been revised to describe more fully the properties of complex (doubly achromatic) magnets used in contemporary treatment units in contrast to the simplified (singly achromatic) magnet shown in Figure 36. A discussion of multileaf collimators has been added to Section 11. Additional technical information on advances in accelerator design may be found in the added reference to *Medical Electron Accelerators* by C. J. Karzmark, Craig Nunan, and Eiji Tanabe, 1993. Appendix A has been revised to include descriptions of contemporary linac treatment units.

Acknowledgments

The need to simplify complex microwave and physics phenomena while retaining rigor in the treatment of these phenomena presented a significant dilemma in writing this primer. We are deeply indebted to our many colleagues who gave generously of their time in critically reviewing the manuscript, suggesting changes, simplifying analogies, and identifying areas that were unclear. Their incisive comments enabled us to have a better perception of how the primer should be written. We also wish to acknowledge the assistance, critical review, and encouragement of BRH staff members Frank Kearly and Marcia Shane. We thank Craig Nunan for important contributions, in particular, Section 9. For recent editorial contributions, we acknowledge J.R. Cameron, Ph.D.

The original work was supported in part by Research Grant CA-05838 from the National Cancer Institute, NIH, and in part by an Interagency Agreement with the National Cancer Institute, NCI 2Y01-10606.

Abstract

Although electron linear accelerators (linacs) are widely used for radiation therapy, few basic training materials on their operation have been produced for use by medical professionals.

This primer promotes an understanding of the safe and effective use of linacs. It describes the major components and explains the operating principles of the linear accelerator and acquaints the reader with pertinent features and terminology.

It is intended for students of radiation therapy, radiological physics, radiation oncology, and radiation control.

1 INTRODUCTION

Cancer patients are treated by radiation, surgery, or chemotherapy. A treatment method proving increasingly effective is radiation, used by itself or in combination with other modalities. The principal radiation modality for the treatment of deep-seated tumors is x-rays of very high energy and penetrating power. Such x-rays are created when high-energy electrons are stopped in a target material such as tungsten. Alternatively, the electrons themselves may be used directly to treat more superficial cancers. The electron linear accelerator (linac) accelerates charged particles in a straight line, in contrast to the circular, spiral, or racetrack orbits that characterize the betatron, cyclotron, and some microtrons. The purpose of this primer is to explain the principles of operation and use of the electron linear accelerator and to acquaint the reader with pertinent features and terminology.

The medical linear accelerator will be introduced by first examining the treatment room. Figure 1 shows a patient being readied for treatment with a

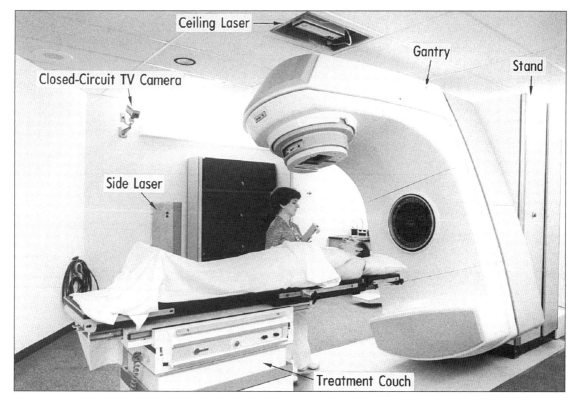

Figure 1. A high-energy radiotherapy linear accelerator. This and many of the illustrations that follow pertain to the Varian Clinac 18.

linac. The thick concrete walls of the treatment room shield the radiation therapist and other staff from the penetrating radiation during treatment. The linac is mounted in a **gantry** that rotates on a **stand** containing electronic and other systems (Figure 2). The linac can be rotated into position about the horizontal gantry axis for use in treatment. The radiation beam emerging from the **collimator** is always directed through and centered on the gantry axis. The beam central axis intersects the gantry axis at a point in space called the **isocenter**. In the majority of cases, the couch is positioned so that the patient's tumor is centered at the isocenter. Usually, the patient lies supine or prone on the **treatment couch** (sometimes called **patient support assembly**). The couch incorporates three linear motions and a rotational motion about the isocenter to facilitate positioning the patient for treatment. Side and ceiling **lasers** project small dots or lines that intersect at the isocenter. These facilitate positioning the patient in conjunc-

tion with reference marks, often tattoos, placed on the patient's skin. **Digital position indicators** display the treatment field size together with collimator and gantry rotation angles. The isocentric system facilitates comfortable, precise reproducible treatment when using multiple fields directed at the tumor from different gantry angles (Figure 3). In this unit, a constant radiation **source-gantry axis-distance** (SAD), usually 100 centimeters (cm), is employed. Alternatively, some treatment techniques use a constant radiation **source-skin** (of patient) **distance** (SSD), usually for large fields at distances of 100 cm or greater.

After positioning the patient for treatment in the treatment room, the radiation therapist confirms the treatment parameters from the **verify and record system** and sets the treatment dose monitor units and time for the patient treatment utilizing the **control console keyboard** (Figure 4). From this position at the control console outside the treatment room, the

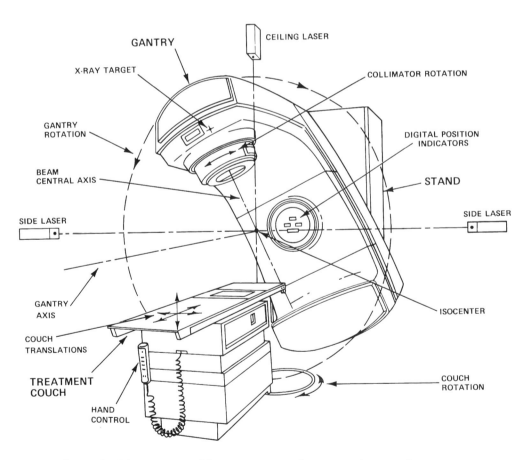

Figure 2. Schematic view of the treatment unit of Fig. 1, emphasizing the geometric relationship of the linac and treatment couch motions.

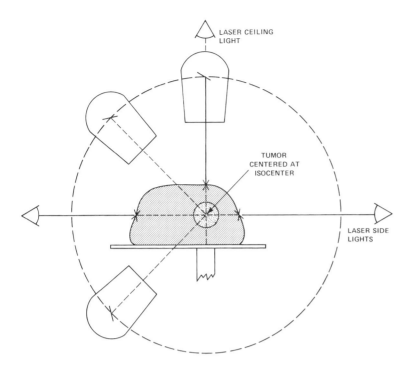

LASER CEILING LIGHT

TUMOR CENTERED AT ISOCENTER

LASER SIDE LIGHTS

Figure 3. The "isocentric" treatment technique. The tumor center, shown within a patient's cross section, is positioned at the isocenter with the aid of skin marks and the lasers shown. The tumor is now positioned for easy and accurate irradiation from any desired gantry angle. The dashed circle depicts all possible x-ray source locations at 100 cm radius (source-axis distance [SAD] = 100 cm).

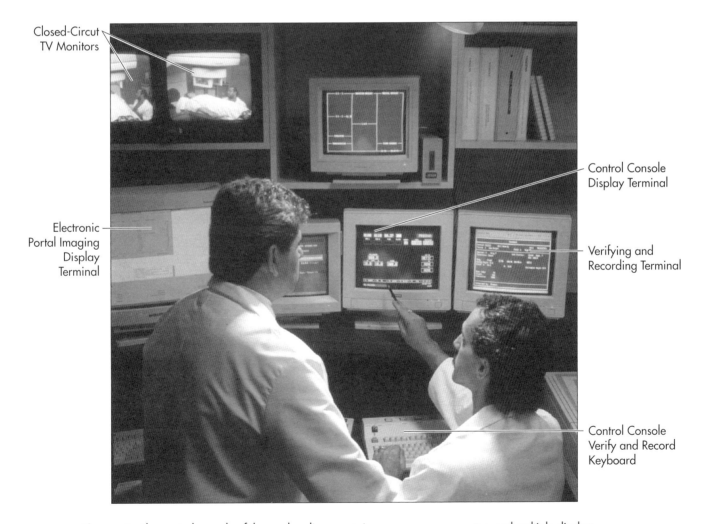

Closed-Circut TV Monitors

Electronic Portal Imaging Display Terminal

Control Console Display Terminal

Verifying and Recording Terminal

Control Console Verify and Record Keyboard

Figure 4. The control console of the modern linac contains one or more computers and multiple display terminals. At the control console, the radiation therapist initiates, monitors, and controls the treatment. The closed circuit television monitors display two views of the patient and the linac. The record and verify system stores, verifies, and displays the patient treatment parameters. The electronic portal imaging terminal displays a real-time image of the photon treatment field. (Courtesy Siemens Medical Systems, Oncology Care Systems).

radiation therapist can view the patient on the closed-circuit TV monitors and the real-time image of the treatment field on the **electronic portal imaging** display and can take emergency action, should it be necessary.

The discussion and illustrations, which follow a brief description of the linac, explain the basic concepts of operation and extend them to the design of an elementary electron linear accelerator. Later, the major modules of a medical linac are identified. Their principles of operation and how they function collectively to produce x-ray and electron treatment beams are described. First, however, we must explain how the energy of a radiotherapy beam is designated.

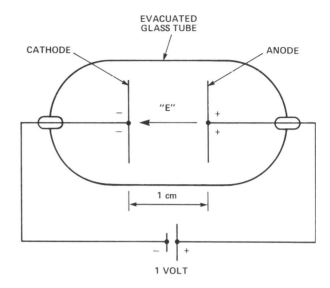

Figure 5. Illustration of an electric "E" field used to accelerate charged particles such as electrons. The "E" field is directed to the left, i.e., from the positive to the negative terminal.

2 ENERGY DESIGNATION IN ACCELERATORS

Figure 5 shows a simple device that will accelerate electrons. It consists of a 1-volt (V) battery connected to two conducting plates spaced 1 cm apart in an evacuated glass tube. The glass tube is an electrical insulator. The negative plate is termed the **cathode** and the positive plate the **anode**. In order to set up the associated electrical charges, the battery causes electrons to flow from the anode to the cathode via the external circuit. This results in a deficiency of electrons at the anode (positive charge) and an excess of electrons at the cathode (negative charge) as shown. This charge distribution creates an **electric "E" field** (denoted by an arrow) in the region between the plates in the direction shown. The size of the electric field is the force that a unit positive charge would feel if placed between the two plates and, in this example, is 1 volt per cm (V/cm). That is, the difference in the electrical potential between the plates, divided by the distance between them, is 1 V/cm. By definition, the arrow identifies the direction a positively charged particle would move; an electron with its negative charge would move in the opposite di-

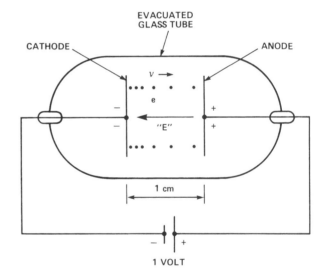

Figure 6. A simple electron linear accelerator of energy one electron volt (1 eV). Electrons, "e," are depicted by dots moving with a velocity (v) to the right. The "E" field is one volt per cm (1 V/cm) in the opposite direction. Note that the density of electrons is highest near the cathode, where their velocity is lowest.

rection. It is not possible to see "E" fields, but they are known to exist because of the force they exert on charged particles such as electrons. If the electrons in Figure 6 are released from the negative plate (the cathode), they will be accelerated by the force of the

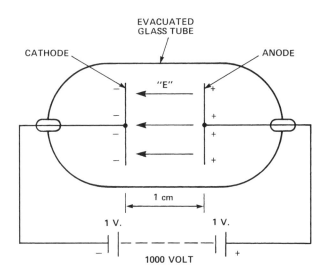

Figure 7. A simple electron linear accelerator of energy 1000 electron volts (1 keV).

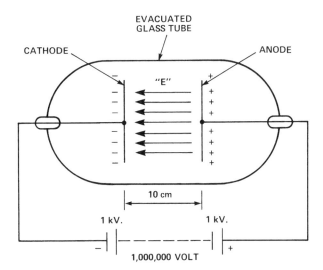

Figure 8. A simple electron linear accelerator of energy one million electron volts (1 MeV). Note the increased electric charge (+, −) distributions at the anode and cathode and greater "E" fields compared with those for the lower voltages of Figs. 5, 6, and 7.

"E" field to the positive plate (the anode). An **electron volt** (eV) is the energy gained by an electron accelerated across a potential difference of 1 V.

Exerting a force through a distance is a basic measure of work and energy. On the atomic scale, the electron volt, or multiples of it, is the adopted unit of energy. In Figure 6 we are dealing with a force of 1 V/cm exerted on an electron through a distance of 1 cm.

Imagine now that a thousand 1-V batteries are connected in series to provide 1,000 V, or 1 kV potential differential, across the plates of this device as in Figure 7. The accelerated electron would arrive at the anode with an energy of 1000 eV or 1 kiloelectron volt (1,000 eV = l keV). Note also that the strength of the associated "E" field now is 1,000 volts per cm (1 kV/cm).

Suppose the plates are spaced 10 cm apart and that a thousand 1-kV batteries are connected in a series to provide a 1-million-V power supply (Figure 8). The plate spacing and glass tube have been lengthened to withstand this higher voltage without electrical breakdown. An electron released from the cathode now gains 1 million eV of energy during its transit and arrives with an energy of 1 million eV (1,000,000 eV = 1,000 keV = 1 MeV). Note that the energy gained by the electron depends only on the potential difference between the anode and the cathode, and not the distance traveled. The corresponding electric field strength "E" is 1 mil-

lion V divided by 10 cm, or 100,000 V/cm (100 kV/cm). To establish the higher electric field strengths of Figures 7 and 8, the + and − charge distributions at the anode and cathode are proportionately larger, as compared with Figures 5 and 6. To simplify the figures that follow, the "E" lines and associated charge distributions at the anode and cathode will sometimes be omitted. Linear accelerators give energy to charged particles by accelerating them in a straight line.

3 AN ELEMENTARY LINEAR ACCELERATOR

It is possible to convert the simple linac just described to a more sophisticated, yet still elementary, electron linear accelerator. First, a heated cathode is substituted for the negative plate (Figure 9). The cathode shown is a simple filament. (In the linac this cathode

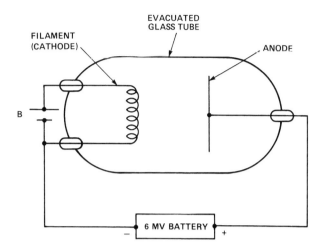

Figure 9. An elementary six million volt (6 MeV) electron linac. The filament-type cathode for the electron source is heated by a small battery, B.

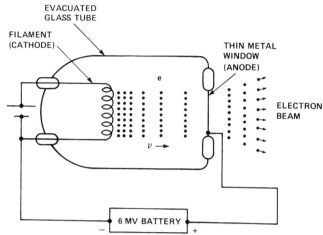

Figure 10. The-six-million-volt linac of Fig. 9 modified to provide external 6-MeV electrons (as for therapy). Electrons, "e," are accelerated and emitted continuously in this simple battery-powered linac.

becomes the electron gun.) The small battery, B, heats the filament causing it to literally "boil off" electrons. Next, a theoretical 6-MV battery is connected between the cathode and anode. This battery voltage corresponds to the electron energy desired, i.e., 6 million V for 6-MeV electrons. Here, electrons are boiled off the filament and accelerated to an energy of 6 MeV as they strike the anode.

To adapt this linac for electron therapy, a **thin metal "window"** becomes the positive plate or anode (Figure 10). Such a thin, solid, metal sheet maintains the necessary vacuum and yet permits the electrons to penetrate the window and emerge with only a small loss of energy. In this elementary linac, the **electron beam** emerges with an energy only slightly less than 6 MeV.

To adapt this linac for x-ray therapy, the positive anode placed outside the window is a thick **tungsten target** that stops the electrons abruptly, thereby producing penetrating x-rays (Figure 11). These x-rays will have energies from a fraction of an MeV up to 6 MeV, all initiated by electrons of 6 MeV energy, since the electrons can give up their energy all at once in a single collision or in several sequential collisions. The resulting spectrum of x-ray energies is designated by "6 MV." The notation convention of dropping the "e" from "MeV" indicates that the x-ray beam will be made up of x-rays of different energies produced as the 6-MeV electrons are slowed and stopped in the target.

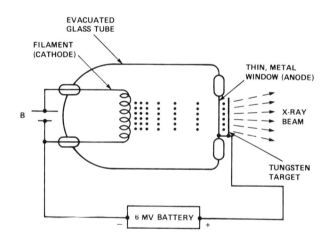

Figure 11. The elementary six-million-volt linac of Fig. 10 modified by providing a target to provide external 6 MV x-rays.

Instead of energizing this simple linear accelerator with a battery, substitute an **alternating voltage**, as shown in Figure 12. The magnitude and polarity of such a voltage changes regularly and repeats itself periodically with time in this cyclic pattern, which is called a sine wave. For the single cycle shown in Figure 12, the horizontal axis denotes time; the vertical axis denotes the magnitude and polarity of the anode voltage, V, relative to the cathode, that establishes the "E" field. Many electrical and mechanical phenomena change smoothly in this regular pattern of a

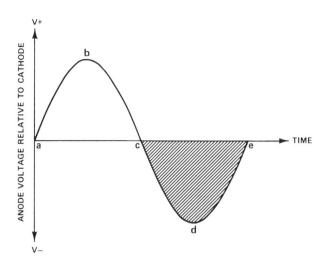

Figure 12. One cycle of the alternating voltage used to power the linac shown in Figs. 13 and 14. The anode voltage, V, relative to the cathode is plotted as the ordinate against time, as an abscissa. This pattern repeats itself at the frequency of the alternating voltage (60 cycles per second or 60 Hz for this elementary linac) as time progresses and is called a sine wave.

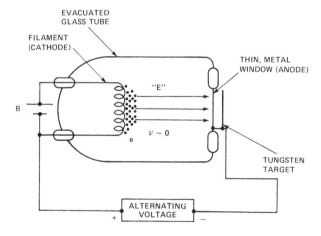

Figure 14. The elementary linac for x-ray therapy powered by an alternating voltage. For the polarity shown (opposite of Fig. 13), electrons remain near the filament and will not be accelerated to the anode.

sine wave. The number of complete sine wave cycles per second (+ and – for up and down excursions) is called the **frequency** and is measured in **hertz (Hz)**, **kilohertz (kHz)**, or **megahertz (MHz)**. One hertz equals one cycle per second. Typically, the frequency of home electric power is 60 Hz, a standard (AM) broadcast radio wave can be 1000 kHz, and FM radio

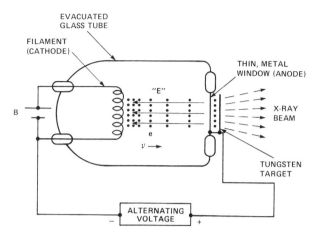

Figure 13. The elementary electron linac for x-ray therapy modified to use an alternating voltage power source, whose polarity reverses every half cycle (see Fig. 12). For the polarity shown, electrons will be accelerated to the anode. Note that "E" and the electron velocity (v) are in opposite directions.

100 MHz. The energizing frequency for medical linacs is 3000 MHz. The latter high frequency is referred to as a **microwave frequency**. The time for completing a single cycle is called the **period** and, for the above examples, coincides with 1/60th second, 1/1,000,000th second (1 microsecond), and, lastly, 1/3,000th of a microsecond, respectively.

Most linacs are powered by a 60 Hz line voltage (see Figure 12), whereas the electrons are accelerated by a 3000 MHz voltage. With the target positive and the filament negative, as shown in Figure 13, electrons emitted from the cathode during interval a-b-c of Figure 12 are accelerated to the target. At point c in Figure 12, the voltage reverses polarity and "E" field direction. With the target negative and filament positive, during interval c-d-e as shown in Figure 14, electrons are still emitted and form a "cloud" of electrons around the filament but are **not** accelerated to the target.

Electrons are accelerated only during the first half of the cycle shown in Figure 12. At a time one-fourth through the cycle, point b, the voltage, V, reaches a positive maximum, and "E" is directed as in Figure 13. An electron released from the cathode at this time would gain a maximum of energy. Conversely, at three-fourth's time through the cycle, point d, V reaches a negative maximum. Then, "E" is directed as in Figure 14, and maximum energy would have to be expended in "pushing" the electron from the cathode if it were to reach the anode against the opposing "E" field. Therefore, no electrons are accelerated at this time. At other times, intermediate amounts of energy would be gained or expended, including zero at points a, c, and e. In this illustration, we assume that the electron travels between cathode and anode instantly; that is, the electron's travel time is zero. Note that now the elementary linac accelerates electrons and emits radiation only half of the time, and the electrons vary in energy sinusoidally during this time.

4 A COMPARISON OF LINACS WITH DIAGNOSTIC X-RAY GENERATORS

There are many similarities between the linac and a conventional diagnostic x-ray generator. Both provide a source of electrons from a hot filament or cathode in an evacuated tube. Both require an accelerating voltage between the cathode and target anode. This voltage is adjustable in a diagnostic generator, depending on the procedure, from about 30 to 150 kV. In contrast, linac accelerating voltages are fixed in a particular unit and range from about 4 to 35 MV. Diagnostic x-rays often involve a single 0.01- to 10-second pulse with a 60 Hz structure with higher frequency ripple components, while linac radiation consists of short bursts of about five-millionths of a second duration repeated several hundred times per second, each burst having a 3000-MHz frequency structure. Both employ collimators to shape the x-ray beam, but these must be thicker in the case of linacs. Because of their high energy, x-rays from linacs are much more penetrating than diagnostic x-rays. This is a distinct advantage for treating a deep-lying cancer because the cancer cells can be destroyed by the linac beam with less damage to healthy, overlying tissues.

Linacs require heavily shielded rooms to protect the persons outside. Such rooms are usually constructed with thick concrete walls. In contrast, diag-nostic rooms are usually shielded by a sheet of lead a few millimeters thick hidden in the walls. Diagnostic x-rays reveal anatomical structures based on differences in atomic number as well as physical density, e.g., bone versus soft tissue or air; megavoltage x-ray attenuation is primarily based on density differences. A film produced with megavoltage x-rays would show little difference between bone and soft tissue. The importance of x-ray diagnostic beams is primarily in the information contained in the transmitted beam, which produces an image on a receptor. The importance of x-ray therapy beams, such as provided by linacs, is primarily in the energy absorbed in the tumor, not in the beam transmitted through the patient.

Orthovoltage (about 250 kV) radiation equipment, which dominated treatment energies of the 1930s, has properties closer to diagnostic x-rays than megavoltage energy therapy beams and continues to be appropriate for specific treatments.

5 MAJOR LINAC MODULES AND COMPONENTS

Contemporary linacs consist of a number of major modules and components that will be identified and their operating principles described. This information will be combined and interrelated to explain an operational linac.

The major modules in the linac are the gantry, the stand, the control console, and the treatment couch (Figures 1 and 4). Some linacs also have an external modulator cabinet, as Figure 43 shows. Figure 15 identifies the components housed in the stand and gantry of a high-energy linac and will be referred to frequently. The stand is anchored firmly to the floor, and the gantry rotates on bearings in the stand. The operational accelerator structure, housed in the gantry, rotates about a horizontal axis fixed by the stand. For ease of understanding, most of the text will describe the components as they appear in the Var-

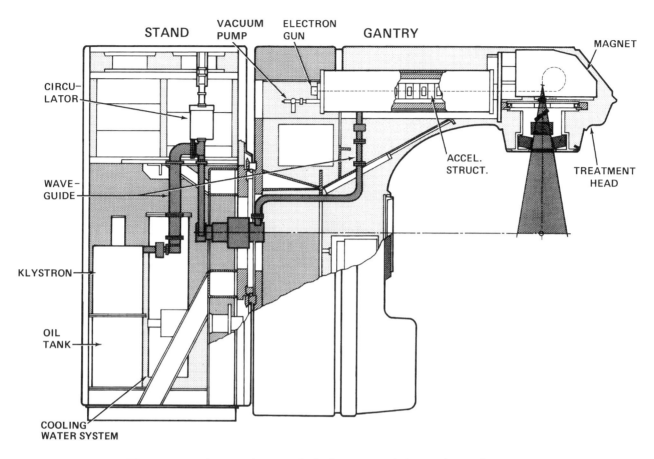

Figure 15. Schematic diagram of a high-energy radiotherapy linac (Clinac 18) identifying major components housed in the stand and gantry.

ian Clinac 18. Variations in linac design do occur and are described in Appendix A.

The major components in the stand are the following:

1. **Klystron**—a linear series of microwave cavities that sit atop an insulating oil tank and provides a source of microwave power to accelerate electrons;
2. **Waveguide**—which conveys this power to the accelerator structure in the gantry;
3. **Circulator**—a device inserted in the waveguide to isolate the klystron from microwaves reflected back from the accelerator;
4. **Cooling water system**—which cools various components that dissipate energy as heat and establishes a stable operating temperature for the accelerator structure that is sufficiently above room temperature to prevent condensation of moisture from the air.

The major components found in the gantry are the following:

1. **Accelerator structure**—a linear series of microwave cavities that are energized by the microwave power supplied from the klystron via the waveguide;
2. **Electron gun** (or cathode)—which provides the source of electrons injected into the structure;
3. **Bending magnet**—which deflects the electrons emerging from the accelerator structure around a loop focusing the electron beam on the target to produce x-rays or to be used directly for electron treatments;
4. **Treatment head**—which contains beam shaping and monitoring devices;
5. **Beam stopper**—which reduces room shielding requirements for the treatment beam emerging from the patient, and extends from the bottom of the gantry as shown in Figure 42a.

The **modulator cabinet** (Figure 43) contains components that distribute and control primary electrical power to all areas of the machine from the utility connection and also supplies high-voltage pulses for beam injection and for generating microwave power.

The **treatment couch** motions are controlled by a hand pendant control operated by the radiation therapist (Figure 1). The three-dimensional positioning of the patient on the couch is motor-driven. Fast and slow speeds or variable speed motor control are provided for the couch, together with control of gantry rotation and secondary collimator adjustments (Figure 2). Most couches also provide couch rotation around a vertical axis passing through the isocenter, and some permit attachment of a treatment chair.

The **control console** (Figure 4) is the operations center for a linac. It supplies the timing pulse that initiates each pulse of radiation. It provides visual and electronic monitors for a host of linac operating parameters including the individual patient's treatment dose. The prescription is for the total series of treatments. Treatment cannot proceed when the treatment parameters exceed limits that have been previously established.

In addition to these major modules and components, there are a number of auxiliary systems including **vacuum and water pressure, temperature control, automatic frequency control (AFC),** as well as radiation **monitor** and **control** (see Section 13).

6 MICROWAVE POWER SOURCES

The **klystron** and **magnetron** are two special types of evacuated electron tubes that are used to provide microwave power to accelerate electrons. Klystrons are used to power high-energy linacs; magnetrons to power lower energy linacs. Microwaves are similar to ordinary radio waves but have frequencies thousands of times higher. The microwave frequency needed for linac operation is 3 billion cycles per second (3000 MHz). See sections 3 and 4. The voltage and "E" fields associated with microwaves change sinusoidally in direction and magnitude in a regular manner, producing an alternating voltage as shown in Figure 12. Microwave cavities, which are central to the construction and operation of klystrons and magnetrons as well as linear accelerator structures, will be described next.

6a. Microwave Cavities

Microwave devices, including klystrons, magnetrons, and accelerator structures, make extensive use of resonant microwave cavities. A simple microwave cavity similar to that used in medical linacs, but with closed ends, is shown in Figures 16 and 17. It is an accurately machined cylinder, about 10 cm in diameter and several centimeters in length. Such a cavity has the approximate size and shape of a 7-oz. tuna fish can. In Figure 18, the cavity is shown modified by cutting openings in its two ends along the axis, for use in a klystron or an accelerator structure. A microwave cavity is an enormously efficient device in the sense that the intense "E" fields needed for these applications are established by a small amount of electrical power. This is a resonance phenomenon that occurs at one frequency, in this case 3000 MHz, which is determined by the dimensions of the cavity much as a musical organ pipe of a particular length resonates to a particular pitch. Such cavities are formed of copper for high electrical and thermal conductivity. An electric current, I, flows on their inner walls, moving electric charge from one cavity end to the other, as shown in Figures 16, 17, and 18. These end regions of dense electric charge are central to both klystron and accelerator structure operation because they give rise to the intense "E" fields along the axis of the cavity, as in Figures 16b, 17b, and 18b. The magnetic "H" field pattern of Figures 16c, 17c, and 18c that exists in the cavity will be omitted in the illustrations that follow, because they are unimportant for our purposes.

The electric and magnetic fields and currents and charge distributions that exist in a cavity have a complex dependence on time. They have been separated arbitrarily in Figures 16, 17, and 18a, b, and c, for clarity. The polarity of the electric charge and current,

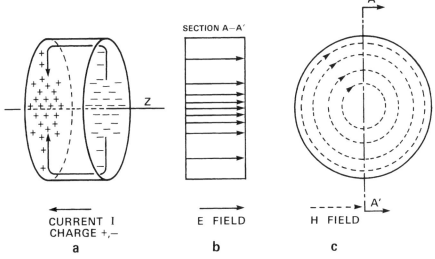

CURRENT I
CHARGE +,−

a

E FIELD

b

H FIELD

c

Figure 16. The electric charge (+, −), current I, and electric "E" field and magnetic "H" field patterns for a closed cylindrical microwave cavity. The cavity wall current, I, is circularly symmetric around the Z axis of the cavity. The "E" and "H" fields fill the entire cavity volume. Section b is made by a plane surface cut through the center of the cylinder containing the Z axis; section c is cut perpendicular to the Z axis.

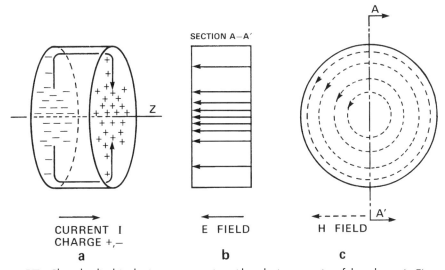

CURRENT I
CHARGE +,−

a

E FIELD

b

H FIELD

c

Figure 17. Closed cylindrical microwave cavity with polarity opposite of that shown in Figure 16.

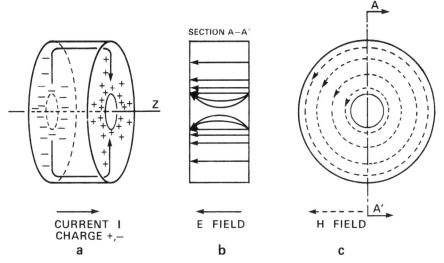

CURRENT I
CHARGE +,−

a

E FIELD

b

H FIELD

c

Figure 18. Cylindrical microwave cavity with the same polarity as Figure 17 but with circular holes cut in the ends of the cavity. The holes concentrate the current, I, charge and the "E" field and facilitate acceleration of electrons along the Z axis.

and the "E" and "H" field directions reverse twice each microwave cycle, that is, 6 billion times a second! The patterns of Figures 16 and 17 are one-half cycle apart in time. In order to take advantage of these intense "E" fields to build a klystron or an accelerating structure, circular openings on axis at the cavity ends are cut as shown in Figure 18 so that electron beams can be introduced to interact with these fields. The electron beam current passes through these openings along the cylindrical axis Z. The large cavity wall currents, I, should not be confused with the smaller axial electron beam current, which originates from an electron gun in a klystron or in an accelerator structure. The arrows denoting I in Figures 16a, 17a, and 18a point in the direction that a positive charge current would flow. (Note that the axial electron beam current is not shown in these figures.) The negative electrons, which are the charge carriers in the pulse, flow in the opposite direction.

Earlier, the energy transfer from a static, and then an alternating, electric "E" field to an electron trans-

ported between two conducting plates was explained. Recall that in one direction of the "E" field, energy is transferred from the "E" field to the accelerating electron. An electron traveling at high speed in the reverse direction can transfer energy from the decelerating electron to the "E" field. This latter phenomenon will be examined in more detail. It provides the basis for generating microwave power by both the klystron and magnetron.

6b. The Klystron

The elementary **klystron**, depicted in Figure 19, is a **microwave amplifier tube** that makes use of two cavities of the type illustrated in Figure 18. The cross-sectional drawing shown in Figure 19 is a view that contains the cylindrical Z axis of the cavities similar to the view shown in Figure 18b. On the left is the cathode, the source of electrons for the klystron, which is given a negative pulse of voltage. This ac-

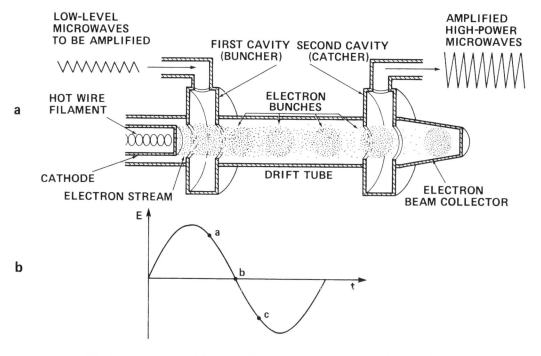

Figure 19. (a) Cross-sectional drawing of an elementary two-cavity klystron tube used as a microwave power amplifier. The two cavities are shown in cutaway sections similar to the section shown in Fig. 18b. A drift tube connects the two cavities. The anode is not a single identifiable component of the klystron but consists largely of the electron beam collector on the right end. (b) The timing diagram is for the "E" field of the first or buncher cavity. The "E" fields vary sinusoidally across the cavity gaps as illustrated for increasing time, t.

celerates electrons into the first, or **buncher cavity**, as it is called. The buncher cavity is energized by very low-power microwaves that set up alternating "E" fields across the gap between left and right cavity walls. The "E" fields vary in time, as shown in Figure 19b. Recall that it is the negative "E" field that accelerates the electrons. Those electrons that arrive early in the microwave cycle, at times between points a and b, encounter a retarding "E" field and are slowed. The velocity of those electrons arriving at time b, when the "E" field is zero, is not affected. Electrons arriving at later times, between points b

and c, are speeded up by the negative "E" field. This causes the electron stream to form bunches. This process is called **velocity modulation**, since it alters the velocity but not the average number of electrons in the beam. They pass along the drift tube connecting the two cavities. The electrons moving with different velocities merge into discrete bunches as shown (Figure 19).

The second, or **catcher cavity**, has a resonance at the arrival frequency of the bunches. As the electron bunches leave the drift tube and traverse the catcher cavity gap, they generate a retarding "E" field by inducing charges on the ends of the cavity and thereby initiate an energy conversion process. By this process, much of the electron's kinetic energy of motion is converted to intense "E" fields in the second cavity, creating microwave power that is used to energize the linear accelerator structure. The residual beam energy that is not converted to microwave power is dissipated as heat in the electron beam **collector** on the far right. The heat is removed by the water cooling system. The beam collector of high-powered klystrons is shielded with lead to attenuate hazardous x-rays created by these stopped electrons. Such klystrons have three to five cavities and are used with high-energy linacs, e.g., 18 MeV and above. The additional cavities improve high current bunching and increase microwave power amplification. They can provide a tremendous (e.g., 100,000 : 1) amplification of microwave power. The klystron is located in the stand, as shown in Figure 15, or in the modulator cabinet for some manufacturers.

Figure 20 illustrates a three-dimensional cutaway high-power klystron that produces about 5 megawatts of peak power and is similar to that used for the Clinac 18. The electron gun (cathode) of the tube is at the bottom. The center section contains four amplifying cavities separated by drift tubes; the upper section consists of the water-cooled collector and output waveguide. This klystron is about 1 meter (m) in length and sits atop an oil-filled tank with its cathode-electron gun portion submerged to provide the requisite electrical insulation (see Figure 15). The cathode is pulsed with a negative voltage of about 120 kV. The four cavities each have tuning adjustments (Figure 20b) that provide small changes in cavity dimensions, bringing them to the same resonant frequency of the accelerator structure operation (some klystrons are pretuned at the factory). The

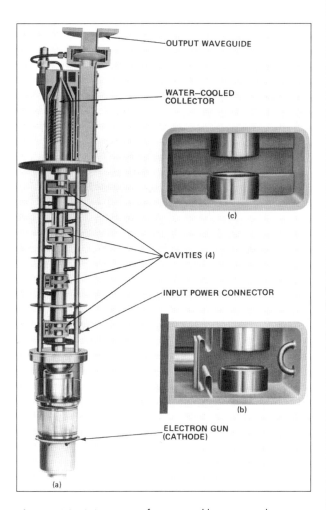

Figure 20. (a) Cutaway four-cavity klystron, similar to that employed in the Clinac 18. Views (b) and (c) are cutaway individual cavity sections. (b) Enlarged view of the bottom cavity, the input power coupling loop is on the right and a fine-tuning device is on the left. (c) Enlarged view of cavity number three; the fine-tuning device has been cut away in this view.

buncher cavity nearest the cathode (Figures 20a and b) is energized from a low-power microwave source. Cylindrical current carrying coils, not shown here, surround the cavities and drift tubes and provide a magnetic field to confine and focus the electron beam traversing the klystron along its axis. The rectangular waveguide conducts the microwave power pulses out of the tube from the output cavity to the accelerating structure.

6c. The Magnetron

The **magnetron** is the microwave source (often called an oscillator) usually employed to power lower energy linacs, typically 12 MeV or less, but occasionally as high as 20 MeV. Like the klystron, it is a two-element tube with a cathode and anode. Both incorporate microwave cavities. The magnetron is usually a less costly (but a less stable) microwave power source than the klystron. The magnetron shown in Figure 21 has cylindrical geometry, shown in the circular cross section of Figure 22. The central cylindrical cathode is surrounded by the evacuated drift

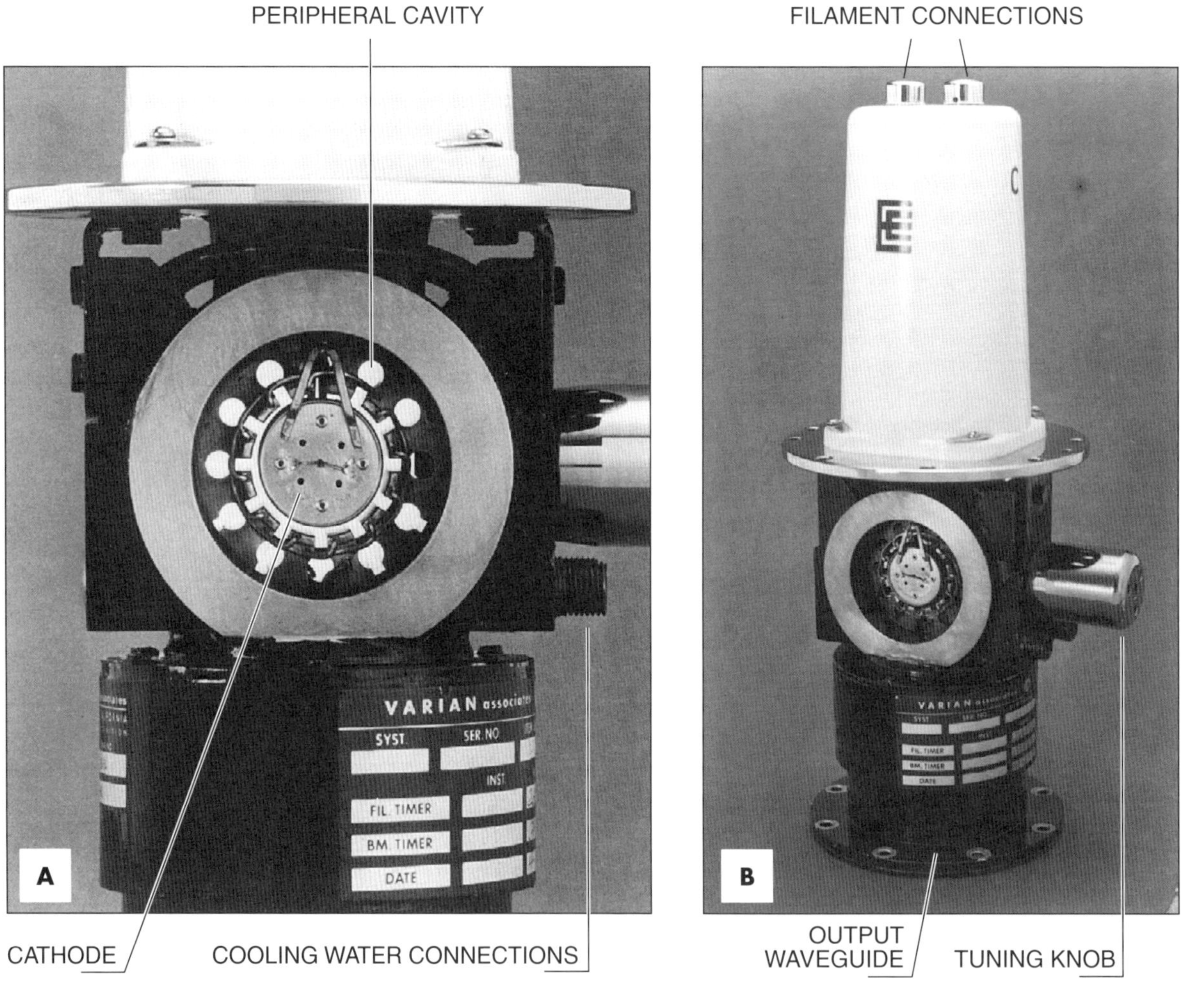

PERIPHERAL CAVITY   FILAMENT CONNECTIONS

CATHODE   COOLING WATER CONNECTIONS   OUTPUT WAVEGUIDE   TUNING KNOB

Figure 21. Cutaway magnetron of a type widely used in medical linacs. (a) The cylindrical cathode is surrounded by 12 peripheral cavities of the segmented anode. (The cavity on the top is obscured by the filament lead used for heating the cathode.) Two small coupling loops, just visible in the bottom cavity, connect the microwave power to the output waveguide just below it. The cooling water connections are on the right. (b) The two filament connections for heating the cathode are on the top; the output waveguide is on the bottom. A fine-tuning knob is on the right.

space and then by an outer anode having 12 cavities. The cylindrical cathode is heated by an inner filament connected to each end of the cylinder, one end of which can be seen in Figures 21a and b. Circular geometry is characteristic of the magnetron; linear geometry is characteristic of the klystron. (Compare Figures 19 and 20 with Figures 21 and 22.)

Figure 22 is a cross section made by cutting a slice at mid-depth, parallel to the surface shown in Figure 21a. A static magnetic field, "H," is applied perpendicular to the plane of the cross section shown. In addition, a pulsed electric field, E_p, directed radially inward all around, is applied between the central cathode and the segmented anode that includes 12 cavities in the outer circular wall.

The electrons emitted from the cathode are accelerated by the pulsed electric field, E_p, toward the anode across the evacuated drift space between cathode and anode. The accelerated electrons induce an additional (+, –) charge distribution shown on the anode poles and an electric field, E_m, of microwave frequency between adjacent segments of the anode (see Figure 22) in a manner similar to that in the catcher cavity of the klystron. In addition, the magnetic field, "H," imparts a circular arc component to the electrons' motion. Thus, they move in complex spirals, S, under the combined influence of E_p; the magnetic field, "H"; and the induced microwave electric field, E_m. In the process, approximately 60% of the kinetic energy of the electron beam is

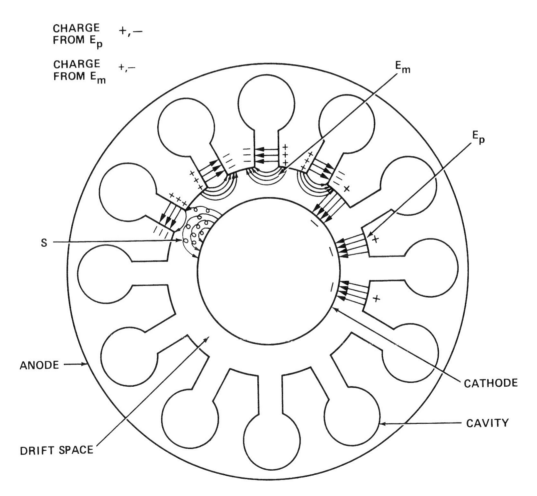

Figure 22. Cross-sectional drawing of the magnetron of Fig. 21 showing representative electric fields E_p (pulsed) and E_m (microwave), with associated electric charge (+, –) distributions. At a particular time and place in the drift space, electrons move in typical paths, S, under the influence of the magnetic field that is perpendicular to the cross section and the sum of electric fields E_p and E_m, which are shown separately.

converted into microwave energy. Magnetrons almost invariably function as high-power oscillators, that is, originators of microwave power, but klystrons usually operate as amplifiers driven by a low-power oscillator. However, if one feeds back a small portion of the output of a klystron to its first cavity, it can function as an oscillator.

The power output of magnetrons and klystrons is measured in thousands of watts (kW) or millions of watts (MW). The **watt** is the unit of electrical power, that is, the rate at which electrical energy is expended. A household electric iron or toaster consumes about 1 kW of electrical power. Typically, magnetrons that operate at a frequency of 3000 MHz (corresponding to a 10-cm wavelength) provide 2 MW peak power output during a burst of radiation, although 4- to 5-MW versions are available at increased cost. The magnetron need only be energized for one one-thousandth of the time to provide the usual short bursts of radiation. Thus, the magnetron shown in Figure 21 operates at 2 MW peak output power and 2 kW average power output and is widely used in medical linacs. Typically, a klystron operates at 3- to

7-MW peak output power and a 3- to 7-KW average output power with an efficiency of about 50%.

7 THE WAVEGUIDE AND CIRCULATOR

Microwave power is conveyed from the klystron (or magnetron) to the accelerator structure by a system of hollow pipes called waveguides (Figure 15). These are either rectangular or circular in cross section, as shown in Figure 23. For example, the short, circular waveguide between the stand and gantry (Figure 15) facilitates rotation of the gantry. The short circular section is between rectangular waveguides. Wave-

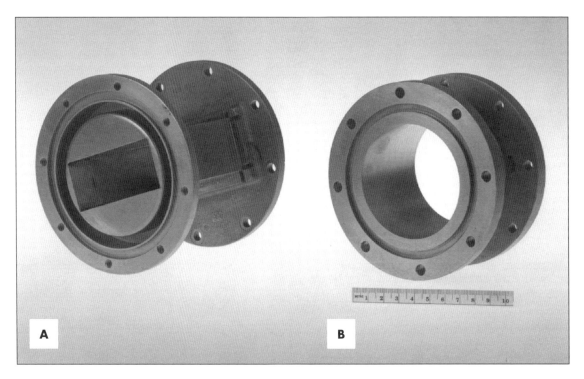

Figure 23. (a) Rectangular waveguide approximately 3.5 x 7.5 cm in cross section; (b) cylindrical waveguide approximately 7.5 cm in inside diameter. (Courtesy of Stanford Linear Accelerator Center, Stanford, California.)

guides replace the traditional electrical wires and cables, which are inefficient in transmitting power at microwave frequencies. Waveguides confine microwaves by reflecting them forward off the walls like a hose or pipe confines water flowing through it. They are pressurized with an insulating gas, sulfur hexafluoride, which reduces the possibility of electrical breakdown and thus increases their power-handling capacity. Two ceramic windows separate the pressurized waveguide from the evacuated klystron at one end and from the evacuated accelerator structure at the other end. The windows are transparent to the microwaves.

The circulator (Figure 15), placed in the waveguide between the klystron and the accelerating structure, acts like a one-way street sign, permitting traffic to move forward through a street intersection but allowing two-way traffic just beyond the intersection. Reflected microwave power is diverted aside in the circulator and absorbed, similar to the way traffic approaching the intersection from just beyond it can be diverted to a side street. Microwave power is allowed to proceed forward from the klystron and through the circulator to the accelerator; but microwave power that is reflected back from the accelerator structure is prevented from reaching the klystron (or magnetron) where it could lead to instabilities and damage.

8 ACCELERATOR STRUCTURES

The linac **accelerator structure** (sometimes called the **accelerator waveguide**) is a long series of adjacent, cylindrical, evacuated microwave cavities located in the gantry, as shown in Figures 15 and 43. It makes use of the cavity principles that have been discussed in Section 6a. Here, however, the objective is to transfer energy from the cavity "E" fields to an accelerat-

ing electron beam. Medical accelerator structures vary in length from 30 cm for a 4-MeV unit to more than 1 m for the high-energy units.

The first few cavities vary in size. They both **accelerate** and **bunch** the electrons in a manner like that of our klystron buncher cavity described earlier. Typically, only about one-third of the injected electrons are captured and accelerated by the microwave "E" field. As the electrons gain energy, they travel faster and faster until they reach relativistic velocity, almost the speed of light. The first few cavities are designed to propagate an "E" field with an increasing velocity in order and to further bunch and accelerate the electrons to relativistic velocity. Later cavities are uniform in size and provide a constant velocity traveling wave, at the velocity of light. Initially, electrons gain energy predominantly by increasing their velocity, later by an increase in their relativistic mass because their velocity cannot attain the speed of light. For example, a 2-MeV electron moves at 98% the speed of light. Its mass in motion is almost five times its mass at rest. Here, we are invoking Einstein's famous mass-energy equivalence concept, that is, increased energy of rapidly moving particles appears as increased mass.

Accelerator structures are of two types: **traveling-wave** and **standing-wave**. The "E" field patterns behave differently in these structures and are central to understanding linacs. First, we will look at traveling wave linacs from this "E" field viewpoint.

8a. Traveling-Wave Accelerator Structures

A hollow, cylindrical pipe, such as the waveguide used for microwave power transmission in Figure 23b, has an "E" field pattern as shown in the cross section in Figure 24a. This pattern travels one way down the pipe from the klystron (or magnetron) faster than the electrons can keep up. Hence, a cylindrical pipe is not useful for accelerating electrons.

The traveling-wave fields are slowed by "loading" the pipe with washer-like inserts called disks (Figure 24b). Now the waveguide has been transformed into a long series of resonant cavities. (Compare Figure 24 with Figure 18.) When energized, these very high cavity "E" fields are suitable for electron acceleration (Figure 24) along the axis.

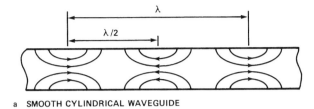

a SMOOTH CYLINDRICAL WAVEGUIDE

b DISK-LOADED CYLINDRICAL WAVEGUIDE

 E FIELD

Figure 24. (a) Spatial traveling-wave electric "E" field pattern at one instant of time along the axis of a smooth cylindrical waveguide. (b) Spatial traveling-wave electric "E" field pattern at one instant of time along the axis of a disk-loaded cylindrical waveguide. The direction of the electric "E" field is reversed every half wavelength, λ/2. The pattern repeats every wavelength, λ, and there are four cavities per wavelength in the disk-loaded structure. The direction of the "E" field also reverses every half cycle in time.

The microwave cavities of the accelerator structures are made of copper. Copper is used because of its high heat conductivity, which improves temperature control, and because of its high electrical conductivity, which reduces power losses. The accelerator structure shown in Figure 25 consists of a series of precisely machined parts, washer-like disks sandwiched between short cylindrical sections. This sequence of disks and short cylinders is assembled on a long spindle for a particular length (energy) structure and soldered together in a furnace. The soldering material is in the form of very thin silver washers, shown at the bottom of Figure 25. These are placed between each disk and cylinder junction surface and, when melted, fuse the components together. Once fused, the sections become a rigid, vacuum-tight accelerator structure. Higher energies require more cavities and longer structures.

The process is still not finished, however. The structure must be tuned to a single, precise, resonant frequency. Machining of the cavity components, prior to assembly on the spindle, is the first step in establishing the correct dimension for each cavity. This amounts to a "rough tuning" and results in a crude resonance with most cavities "off-tune." Next,

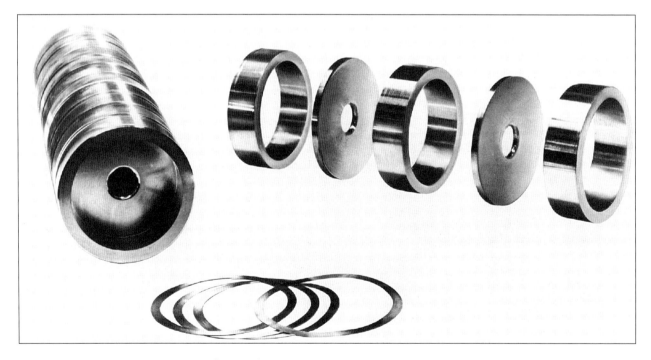

Figure 25. Components of an accelerator structure (right and bottom). Short section of an assembled accelerator structure (left). (Courtesy of Stanford Linear Accelerator Center, Stanford, California.)

each cavity is "fine-tuned" by mechanically squeezing it to create very small dimensional changes, perhaps a few thousandths of a centimeter. Then, like a finely tuned symphony orchestra, they all play the same note. The cavities now resonate to the same frequency and provide optimal energy gain for the accelerating electrons.

As noted earlier, the electrons are captured and bunched on a moving "E" field, gaining energy by traveling in step on the advancing electric wave. At the far end the residual microwave power, not transferred to the electron beam, is absorbed by resistive material fused to the wall of the last cavity, and none is reflected. Further detail of how the wave progresses appears in Figure 26. The "E" field along the axis varies smoothly in a sine wave pattern, as shown for three sequential instants of time, and the pattern

moves smoothly from left to right as time progresses. The solid arrows along the axis denote the instantaneous positions of the maximum positive (to the right) and maximum negative (to the left) values of the traveling electric wave ("E" field). Electrons are accelerated to the right on the negative portions of the "E" wave, that is, just to the right of the sine wave crest identified by arrows directed to the left. In any one cavity, the "E" field maximum reverses direction from time t_1 to t_3 (a half cycle of time) but a given wave crest (direction arrow) travels forward by one cavity from time t_1 to t_2 and again from time t_2 to t_3. From this traveling wave, an electron at a corresponding speed will gain energy in each successive cavity. Figure 27 shows an early prototype traveling-wave accelerator structure, cut in half along its cylindrical axis. Note that the buncher section on the left

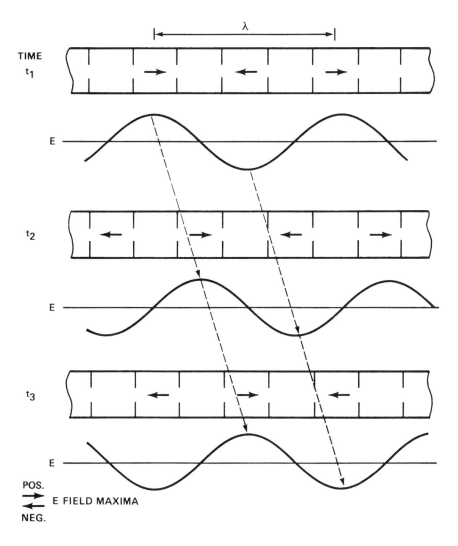

Figure 26. Illustration of a traveling-wave electric "E" field moving to the right in an accelerator structure at three sequential instants of time t_1, t_2, and t_3 separated by one quarter cycle in time. The dashed lines indicate how the "E" field maxima advance to the right with time. There are four cavities per wavelength, λ.

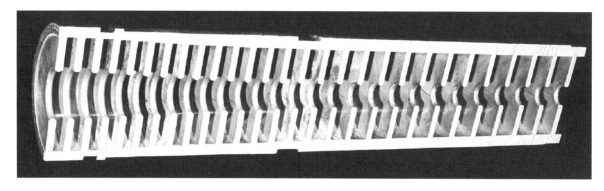

Figure 27. Cutaway traveling-wave accelerator structure; the tapered buncher section is on the left, and the uniform section is on the right.

incorporates larger and variable disk aperture sizes and more closely spaced disks than the uniform section on the right. This buncher is many cavities in length. Bunchers in contemporary linacs use fewer cavities. The input power waveguide attaches on the left (buncher) end.

A boy surfing on a water wave provides a useful traveling-wave analogy in Figure 28a. Here, he is shown riding the forward edge of the crest, moving in step with the wave traveling to the right. If he slips backward over the crest of the wave, he will just bob up and down as the waves pass under him and he will move slowly, if at all, towards the right. Similarly, electrons move forward on the front of the advancing negative "E" wave (Figures 28b and c) or are lost from it. The water particles themselves just go up and down and do not move forward, yet the top of a water wave travels forward. Similarly, the conduction electrons in the cavity walls are confined to moving back and forth between walls of a single cavity, yet the position of the "E" field maximum travels forward as a result of that movement, as shown in Figure 26.

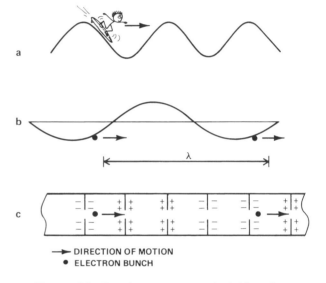

Figure 28. Traveling-wave principle: (a) for a boy surfing on a water wave advancing to the right; (b) for electrons occupying a similar position on an advancing (negative) electric "E" field; and (c) the associated charge distribution that pushes (– charge) and pulls (+ charge) the electron bunches along the cylinder. There are four cavities per wavelength, λ.

8b. Standing-Wave Accelerator Structures

Most present-day medical linacs are of the standing-wave type because the accelerator structure can be much shorter, and, therefore, the treatment unit is less bulky than traveling-wave linacs of comparable energy. Standing-wave linacs operate somewhat like the traveling-wave unit just described but with one significant difference. The "E" wave varies in magnitude with time in a sinusoidal manner, but the pattern remains stationary along the axis and does not advance like the traveling "E" wave or water wave just studied. A good standing-wave analogy is the pattern of a violin string fixed at both ends and vibrating up and down to produce a musical note.

In the case of traveling-wave accelerators, microwave power is fed to the structure via the input waveguide at the proximal (electron gun) end. The residual power is absorbed at the distal (target) end of the structure. In the standing-wave accelerator the microwave power can be fed anywhere along the length of the structure because the power proceeds in both forward and backward directions from the input waveguide and is reflected at both ends. The incident forward wave is reflected backward from the distal end, and the backward wave is reflected forward from the proximal end. There are now two waves: an advancing incident (forward) wave and a reflected (backward) wave. These two waves are reflected back and forth from one end to the other end of the accelerator structure about one hundred times during a 5-microsecond pulse. The circulator, described earlier, stops reflected power from reaching and detuning or damaging the klystron or magnetron.

Figure 29 shows the "E" field maximum values, denoted by arrows for these two waves at three sequential instants in time t_1, t_2, and t_3. The forward wave crests (instantaneous positions denoted by arrows) moving to the right advance one cavity length during the time interval from t_1 to t_2, t_2 to t_3, etc. Sim-

ilarly, the backward wave crests move at the same speed to the left. These sequential movements can be seen by examining each of the two patterns of arrows at the three times. Here, the sine wave "E" field patterns have been omitted and attention is confined to the wave crests denoted by arrows.

The effective "E" field, in accelerating the electron beam, is the sum of the forward and backward waves, as shown in Figure 30. Its magnitude, assuming 100% reflection and no losses, is double that of either the forward or backward wave when the fields in a single cavity are in the same direction. However, it is zero when the fields in a single cavity are in opposite directions. The effective "E" field exhibits a sinusoidal pattern with distance along the accelerator structure, as shown in Figure 30. The crests of the sine wave pattern oscillate up and down with the progression of time.

Note that every other cavity of this standing-wave structure in Figure 30 has a zero "E" field at its center at all times—at times t_1 and t_3 because both the forward and backward "E" fields are zero and at times t_2 because the forward and backward "E" field are equal in magnitude but opposite in direction, and

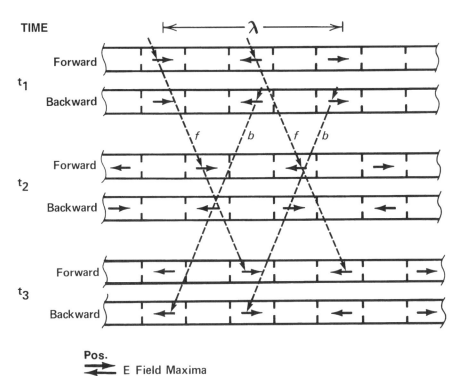

Figure 29. Standing-wave principle illustrating forward (*f* moving to the right) and backward (*b* moving to the left) "E" field maxima at three sequential instants of time. The dashed lines indicate the time sequential positions of the forward and backward moving "E" field maxima. They exist simultaneously in a standing-wave accelerator structure but are shown here in two separate structures for clarity.

cancel completely (see Figure 29). These zero "E" field cavities are essential in transporting microwave power but do not contribute to electron acceleration. Their role is to transfer or couple power between accelerating cavities. Because they play no role in acceleration, they can be moved off-axis and the length of the structure can be shortened.

Figure 31 illustrates how the shortened, side-coupled standing-wave structure evolves from the standing-wave structure of Figure 30. First, every other cavity of Figure 31a, which couples power between accelerating cavities, is shortened in length, as in Figure 31b. The resonant frequency of a cavity depends on its diameter, not its length. Next, they are moved off-axis, as in Figure 31c, and finally, in Figure 31d, placed on alternating sides of the axial accelerating cavities. The spatial "E" field pattern shown

below each sequential accelerator structure is for the same time in the microwave cycle. In Figures 26 and 28b and c, the "E" wave repeats every four cavities, and there are four cavities per wavelength λ. At any given instant, only one of four cavities is accelerating the electron bunch and the other three cavities are "coasting." In Figures 31c and d, the "E" wave repeats every two axial cavities so that, at any instant, half of the axial cavities are accelerating the electron bunch, and the small relatively lossless off-axis coupling cavities replace half of the cavities of the traveling-wave accelerator, hence the shorter length and greater efficiency for the standing-wave design. Figure 32 illustrates in detail how the axial "E" spatial pattern changes in time over a complete microwave cycle for a standing-wave linac. Contrast the time variation of this pattern to that for a traveling-wave

Figure 30. Standing-wave electric "E" field patterns in an accelerator structure for combined forward and backward waves at three sequential instants of time. Two traveling waves moving in opposite directions, as shown in Fig. 29 (*f* forward, upper maxima; *b* backward, lower maxima), generate such a standing wave. The pattern shown at time t_1 will recur one half cycle later at time t_3 with polarity reversed.

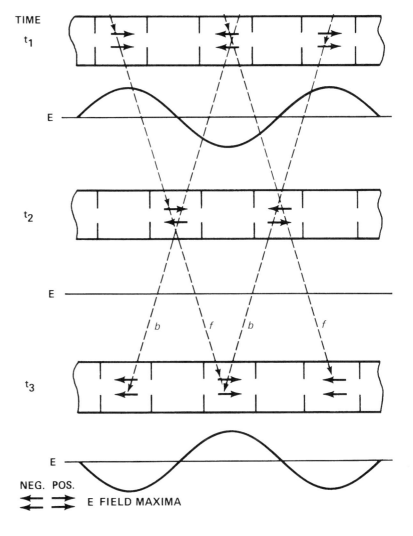

linac in Figure 26. Note that the "E" field pattern does not advance, but changes in magnitude and direction with time. We can now optimize the cavities along the beam axis for acceleration and the off-axis coupling cavities for microwave power transport. Figure 33 is a cutaway view of such an optimized standing-wave accelerator structure. This is called a **bimodal** or **side-coupled** accelerator structure. Two standing-wave ac-

celerator structures constructed in this way are shown in Figures 34 and 35. They are shorter in length than a traveling-wave structure for a given energy gain and a given klystron or magnetron power.

Electrons injected into standing-wave structures, such as those illustrated in Figures 34 and 35, are captured, bunched, and accelerated in the first few cavities, just as in the traveling-wave accelerator. They

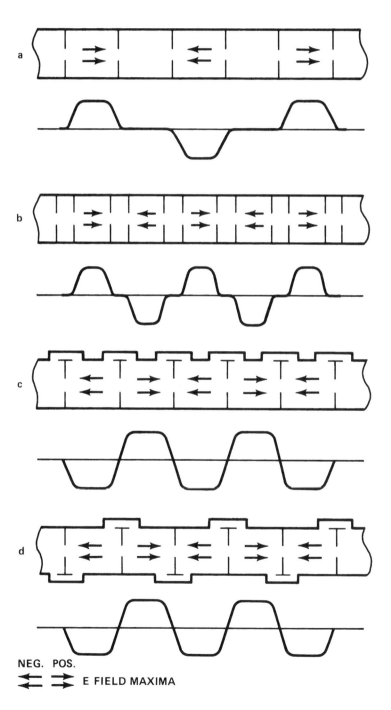

Figure 31. Evolution of side-coupled standing-wave accelerator structure from a linear standing-wave structure and related "E" wave patterns along the axis. The "E" field pattern below each structure shows the spatial field along the axis at the same time in the microwave cycle. The waves are squared due to the presence of harmonics of the fundamental frequency.

NEG. POS.
← ⇉ E FIELD MAXIMA

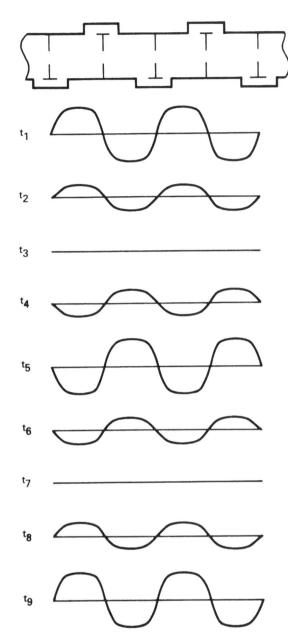

Figure 32. A sequential look at the axial standing-wave "E" field pattern for one full microwave cycle of time for the bimodal structure depicted in Fig. 31d. Note that the "E" field pattern does not advance, but changes in magnitude and direction with time.

Figure 33. Cutaway view of a bimodal or side-coupled standing-wave accelerator structure. The accelerating cavities are shaped for optimum efficiency. The coupling cavities are staggered to reduce asymmetries introduced by the coupling slots. (Courtesy of Los Alamos Scientific Laboratory, Los Alamos, New Mexico.)

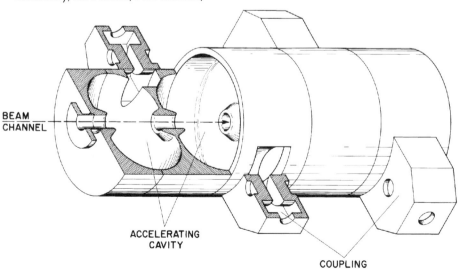

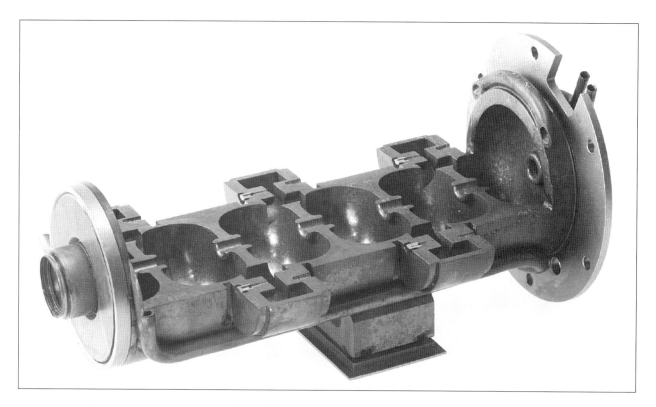

Figure 34. Cutaway view of a standing-wave linac structure. Electrons attain an energy of 4 MeV in this 30 cm long structure with five accelerating cavities. The input waveguide is on the bottom, the electron gun attaches on the left, and the x-ray target is permanently sealed into the structure on the right end. (Clinac 4, courtesy of Varian Associates.)

Figure 35. Linac standing-wave structure. The input waveguide is on the top just right of center. Water-cooling tubes, which are soldered to the structure, can also be seen. (Clinac 18, courtesy of Varian Associates.)

pass through the following cavities during the "E" wave's negative excursion and are accelerated (recall that electrons are accelerated in the opposite direction of "E"). During that time, the "E" wave of the next adjacent cavity is positive and electrons are not accelerated in it. However, as the electron bunch crosses the boundary between adjacent cavities, the "E" wave in the next cavity starts its negative excursion and the electron bunch is again accelerated. Each cavity accelerates electrons only when its "E" field is negative. This process continues until the electrons acquire their final energy.

9 DUAL X-RAY ENERGY MODE ACCELERATORS

In recent years there has been a move toward the use of dual x-ray energy mode linacs in radiotherapy. For example, such a machine might provide a 6-MV x-ray mode, an 18-MV x-ray mode, and an electron mode with energies of 6, 9, 12, 15, and 18 MeV. For patients benefiting from treatment in more than one mode, greater precision of patient positioning can be obtained because the patient does not have to be moved from one treatment couch and treatment room to another. There are also cost savings in having the full range of radiation modes within only one treatment room and machine instead of two, such as for radiotherapy departments having modest patient volume. Dual x-ray energy mode linacs have become available using either a standing-wave or a traveling-wave accelerator structure.

For dual x-ray energy mode linacs, the accelerator structure length must be adequate to provide the high-energy x-ray mode at a high dose rate. This structure must also provide the low-energy x-ray mode at a high dose rate. This latter requirement necessitates a high beam current because x-ray production efficiency and, therefore, dose rate is markedly reduced at the low electron energy. All electron mode energies require much smaller beam currents, which are easily provided.

9a. Standing-Wave Accelerators

Early linacs operated at one low x-ray energy constrained by the use of low-power magnetrons and the need for high beam current for adequate x-ray output (e.g., 1 gray/min at 100 cm SSD) and reasonably short treatment times. Treatment beam energy is increased by higher microwave power input for a given beam current. As accelerator structures and microwave power sources became better understood, variable energy linacs for electron therapy and dual energy linacs for x-ray therapy were developed.

The standing-wave structure for a dual x-ray energy linac can be considered as consisting of two portions, the output end and the gun end. If, in order to reduce x-ray beam energy (e.g., from 18 to 6 MV), the magnitude of the accelerating "E" field is reduced in the second portion (output end) of an unmodified standing-wave accelerator structure, the magnitude of the "E" field drops correspondingly in the first portion (gun end). Similarly, if the phase of the sine wave "E" field is shifted in the second portion, it shifts equally in the first portion. Because the RF power is reflected back and forth in a standing-wave structure, the first portion senses and adjusts to the field in the second portion, and vice versa.

If the magnitude of the "E" field is correct for 18 MV operation—that is, optimal capture, bunching, and positioning of electrons injected from the electron gun on the "E" wave crest and for acceleration through the remainder of the structure—then the magnitude of the "E" field must be reduced for 6-MeV operation. A large beam current must be attained so as to provide a high 6-MV dose rate. In present dual x-ray energy linacs there are two fundamentally different ways to modify standing-wave accelerator structures to eliminate or reduce this problem of inadequate electron capture, bunching, and positioning in the low x-ray energy mode.

1. Change the ratio of microwave power fed to the first and second portions of the standing-wave accelerator structure. This can be done by using a compact **energy switch** in a side cavity located between the first and second portions, as shown

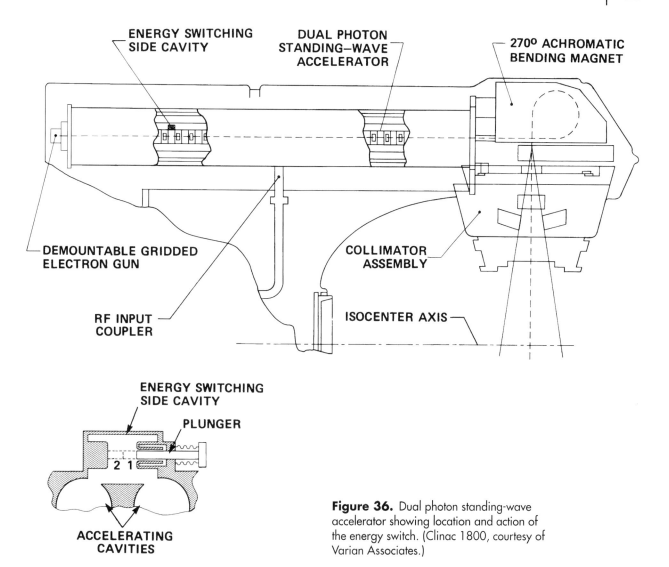

Figure 36. Dual photon standing-wave accelerator showing location and action of the energy switch. (Clinac 1800, courtesy of Varian Associates.)

in Figure 36. In one position of the energy switch, the side cavity provides high coupling between the first and second portions, creating a single high-magnitude "E" field in both portions of the accelerator structure for the high x-ray energy mode. In the second position of the energy switch, the side cavity provides low coupling, creating a low-magnitude "E" field in the second portion of the accelerator structure for the low x-ray energy mode. In both positions of the energy switch, the required "E" field is maintained in the first portion of the accelerator structure in order to maintain optimal capture and bunching of the injected electrons from the gun and positioning the resulting bunch near the crest of the "E" wave. This results in an electron beam of stable energy with a narrow energy spread and, hence, high transmission through the bending magnet and minimal leakage radiation due to electrons lost from the beam before reaching the x-ray target. This provides high dose rate and a stable x-ray beam in both low- and high-energy modes. The energy switch employs one moving part, a plunger. Alternately, instead of an energy switch, a high-power microwave circuit can be used, employing a power divider and a phase shifter. Such systems, which necessitate modification of the design and operation of the structures, are bulky and employ many moving parts.

2. Use a broad band buncher in a standing-wave accelerator structure. In the first portion of such an accelerator structure (the buncher), the cavities are

made very short. Also, the coupling slots to the side cavities may be small to reduce the magnitude of the "E" field in this first portion. There is no energy switch, so the magnitude of the "E" field is one value for high-energy x-ray mode and a much lower value for low-energy x-ray mode, throughout the accelerator structure. This is achieved by reducing the output of the klystron or magnetron for the lower energy mode.

Because the initial cavities are so foreshortened, the electrons injected from the gun are captured and bunched around a position very far forward of the crest of the accelerating "E" field sine wave. Following this bunching section there is one especially long cavity. In passing through this cavity the electron bunch slips backward relative to the "E" field sine wave to near (but not on) its crest, for acceleration through the rest of the accelerator structure; ahead of the crest in a high "E" field for high x-ray energy mode; behind the crest in a low "E" field for low x-ray energy mode.

This technique avoids the use of a mechanically moving part, namely the plunger in the energy switch. However, it is wasteful of RF power, requiring a higher power klystron or magnetron, and the off-crest acceleration produces an output beam with larger energy spread and greater energy instabilities. This makes it more difficult to obtain high transmission of the electron beam through the bending magnet to ensure a high dose rate flattened fully to the corners of large fields in low-energy x-ray mode, with stable dose distribution over all gantry angles.

9b. Traveling-Wave Accelerators

In traveling-wave accelerator structures without RF feedback through an external circuit, the first portion does not sense the field in the second portion because the wave travels only forward. The amplitude of the accelerating "E" field can be changed in the second portion of the accelerator structure without significant effect on the capture and bunching properties of the first portion. One way of producing a downward taper of this "E" field from first to second portions is by beam loading, simply increasing injected beam current from the gun and keeping the klystron or magnetron power constant. Because the

RF power is being transferred to the high-current electron beam, a progressively decreasing fraction of the RF power flowing through the accelerator structure is left to produce the "E" field in the cavities of the second portion. Also, the phase of the "E" field sine wave can be tapered from the first portion to the second portion, simply by changing the frequency of the klystron or magnetron; the electron bunch then slips in phase over the "E" field sine wave, receiving less than maximal acceleration. As pointed out in Section 8b, traveling-wave structures are much longer than standing-wave structures for the same input RF power, beam energy, and beam current. Such long accelerator structures for dual x-ray mode accelerators can be accommodated more easily in a drum-type gantry. Here the accelerator structure can project through the barrel-like drum, which rotates on bearings mounted on the floor.

External RF feedback from the second portion output to the first portion input is used for some traveling-wave structures in order to improve frequency stability (so that a magnetron can be used in a high-energy accelerator), and in this respect the first portion senses the field in the second portion. Variable coupling is used in such feedback circuits to maintain the same accelerating "E" field in the first portion both in high x-ray energy mode with light beam loading and in low x-ray energy mode with heavy beam loading.

The frequency and beam stability of traveling-wave structures and magnetrons are inherently less stable than standing-wave structures and klystrons. Linac designs incorporating them rely more heavily on electronic feedback, such as using computer lookup tables, to maintain treatment beam stability.

10 BENDING MAGNET

The electron beam leaving the accelerator structure continues through an evacuated bending magnet system. Here it is deflected magnetically so as to either strike a target for x-ray therapy or to exit through the treatment head, via a thin metallic window, for electron therapy. The exit axes coincide for x-ray therapy and for electron therapy. Note the location of the bending magnet in Figure 15.

The bending magnet deflects the beam in a loop of approximately 270° (Figures 37 and 38). This magnet design provides focusing for the spread of energies in the beam. Whereas a simple 90° deflection magnet will defocus and spread the beam, a small x-ray focal spot helps ensure that the x-ray treatment fields have sharply defined edges (i.e., a small penumbra), a feature that is of assistance in treatment. This feature improves uniformity of radiation of the tumor and spares nearby critical organs. Medium- and high-energy accelerators employ bending magnets. However, many low-energy units have a straight-through beam without a bending magnet. This is be-

cause these accelerator structures are short enough to be vertically mounted and still allow isocentric rotation up to 6 MV.

As shown in Figure 37, the lower energy electrons are deflected through a loop of smaller radius, and the higher energy electrons are deflected through a loop of larger radius. The important property of this 270° **achromatic magnet** is that these components of energy are brought back together to the same position, angle, and beam cross section at the target, as they were when they left the accelerator structure. This achromatic focusing property is analogous to an achromatic camera lens wherein the different colors (wavelengths) of light from the object are focused to the image on the film. Thus, a 3-mm diameter beam out of the accelerator is reproduced as a 3-mm diameter beam at the target.

However, in the singly achromatic magnet of Figure 37, a variation in mean beam energy will

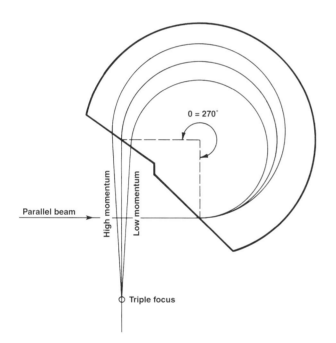

Figure 37. A simplified 270° beam-bending magnet (singly achromatic) with focusing properties as shown. The magnetic field, H, is perpendicular to the plane of the electron orbits. A full doubly achromatic magnet includes additional angular and spatial focusing properties not shown here but described in the text. For example, it also provides transverse focus, that is, focusing in a plane at right angles to that shown. (Courtesy of Physics and Medicine in Biology, Vol. 18, pp.321–354, 1973 and C.J. Karzmark, Ph.D.)

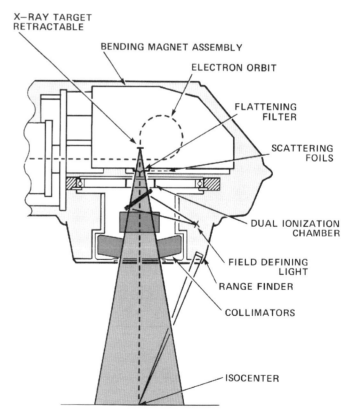

Figure 38. Treatment head. (Clinac 18, courtesy of Varian Associates.)

result in a change in the angle of the beam at the target, producing angular x-ray field asymmetry, even though the focal point position stays fixed. To correct for this angular dependence on energy, modern magnets are doubly achromatic. The mean energy of the beam can vary without changing the mean position or angle of the beam at the x-ray target, hence, maintaining symmetry of the treatment field.

11 TREATMENT HEAD

The **treatment head** (Figure 38) contains a number of beam-shaping, localizing, and monitoring devices. The high-energy x-rays emerging from the target are forward-peaked in intensity, being of higher intensity along the beam central axis and of progressively less intensity away from it (see Figure 39). The forward-peaked x-ray lobe is flattened to provide uniform treatment fields. This is accomplished by the **flattening filter**, a conical metal absorber, placed on the axis as shown. It absorbs more photons from the intense central axis and fewer from the periphery of the beam.

The **dual ionization chamber** system samples the radiation beam (x-rays or electrons) passing through the treatment head and produces electrical signals that terminate the treatment when the prescribed dose is given. Two independent ionization chamber channels ensure that the prescribed dose is delivered accurately and safely, with one serving as a check on the other.

The **field defining light** simulates the x-ray field by illuminating the area to be irradiated on the patient's skin surface and facilitates positioning the patient for x-ray treatment. It provides an intense light field, duplicating in size and shape the x-ray field incident on the patient as defined by the collimators or other beam-limiting devices such as beam blocks or apertures.

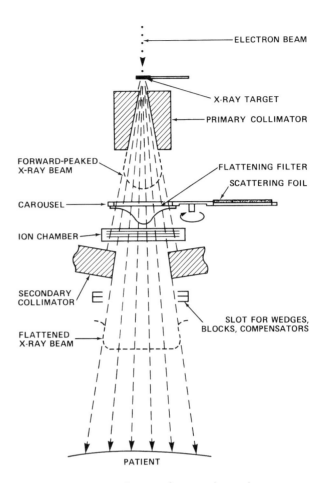

Figure 39. Beam subsystem for x-ray beam therapy.

A **range finder** (optical distance indicator) light projects a numerical scale on the patient's skin to indicate the source-skin distance (SSD) from 80 to 130 cm on the central axis.

The **x-ray target** and **field flattener** are retractable and are moved off-axis for electron therapy.

Additional details of the treatment head subsystem for **x-ray therapy** are shown in Figure 39. A **primary collimator** limits the maximum field size for x-ray therapy. The effect of the flattening filter on beam uniformity is shown. The forward-peaked x-ray beam has been flattened. Treatment field size is defined by the **secondary collimator** consisting of four thick metal blocks, called jaws, often made of tungsten. To help provide sharp edges for treatment fields, the movement of the blocks is confined to arcs so that the block faces present a flat edge to the beam diverging from the target. They are adjustable in pairs (or independently in some units to provide asymmetric fields) and, in some linacs, provide rectangular treatment fields as large as 40×40 cm at a distance of 1 m from the target. The secondary collimator rotates about the beam axis, allowing angulation of fields. Accessories to modify the emergent x-ray field externally, such as wedges, tissue compensators, individually shaped apertures, and shadow blocks, may be mounted on trays that slide into slots of an accessory mount attached to the treatment head.

Individual patient tumors vary in size and shape. Traditionally, the rectangle fields provided by conventional collimation have often been augmented by external contoured blocks cut to the required shape for the individual patient and attached externally to the radiation head accessory mount. The **multileaf collimator** (MLC), by dividing one pair of jaws into

small (e.g., 1 cm at isocenter) adjustable segment widths, provides a satisfactory approximation for the external blocks with a significant savings in construction and attachment time as well as cost. The segments (or leaves) are quickly set daily by computer for the individual treatment fields. Most MLCs allow conventional (rectangular) and conformal (shaped) field treatments with a single linac. Figure 40 shows one MLC design that has 52 motorized leaves, each 1 cm wide at isocenter, with 16 cm travel across the beam centerline.

Improved diagnostic treatment planning information provided by computerized tomography (CT) and magnetic resonance imaging (MRI) have given rise to more precise and extensive treatment planning. Additionally, computer automation of linacs

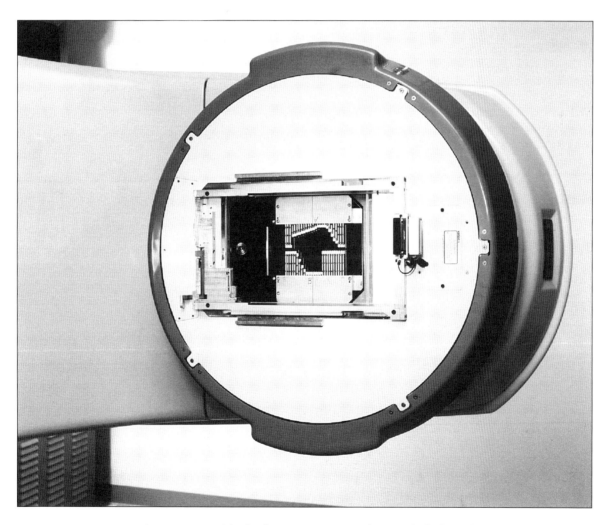

Figure 40. Multileaf collimator (MLC). Note the irregularly shaped field defined by the MLC leaves. (Courtesy of Varian Associates.)

has benefited monitoring and control of treatment. The availability of MLC, together with fast computers, has resulted in dynamic treatment procedures. Here the treatment field size and shape are varied as the gantry rotates around the patient. The changing tumor field contour is provided by the MLC as appropriate. This treatment method is called **dynamic conformal therapy**. As a result, some traditional two-dimensional treatments (and planning) have become three-dimensional procedures.

Additional details of the treatment head subsystem for **electron beam therapy** are shown in Figure 41. The x-ray target is moved out of the beam and a thin **scattering foil** replaces the flattening filter. A rotating carousel or sliding tray(s) facilitates this exchange. The scattering foil spreads out the small, pencil-like beam of electrons and provides a flat uniform electron treatment field. For electron therapy a **detachable electron applicator** is attached to the **accessory mount** of the treatment head. Final field definition is provided by a removable aperture located at the end of the applicator close to or in contact with the patient's skin. In addition, the secondary x-ray collimator is set to a field size somewhat larger than that defined by the applicator. In some linacs the small, pencil-like beam of electrons emerging from the accelerating structure is scanned in a television-like raster pattern to achieve uniformity over the electron treatment field.

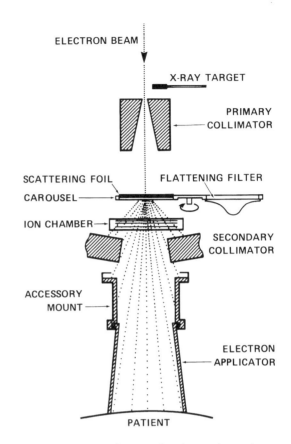

Figure 41. Beam subsystem for electron beam therapy.

12 RETRACTABLE BEAM STOPPER

In most cases the major portion of the treatment beam is absorbed in the patient; the remainder continues on through. This exit beam, which emerges from the patient, spreads out (to widths of a few meters) on the walls, ceiling, and floor of the treatment room. Concrete barriers approximately 2 m thick are needed to reduce 10-MV x-ray intensity and protect personnel outside the room from these direct beams. The wall thickness exposed directly by the beam can be reduced significantly by using a beam stopper (Figure 42a). The beam stopper, constructed of steel and concrete, absorbs 99.9% of the incident radiation. As a result, only the leakage and scatter radiation need be shielded, and a concrete barrier of more uniform thickness (approximately 1 m) for all walls will then be sufficient, thus simplifying room construction and also saving space. Although the use of a beam stopper reduces barrier thickness requirements, access to the patient being readied for treatment is more restricted unless it can be retracted, as in Figure 42b. The beam stopper is fully extended by motor control prior to treatment and is interlocked to prevent treatment when it is not in position. The treatment unit illustrated in Figures 1, 2, and 15 incorporates a counterweight instead of a beam stopper and would be installed in a room of sufficient wall thickness to protect personnel.

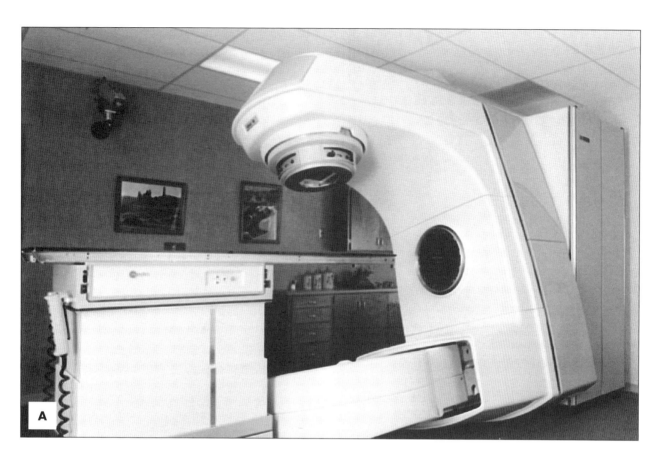

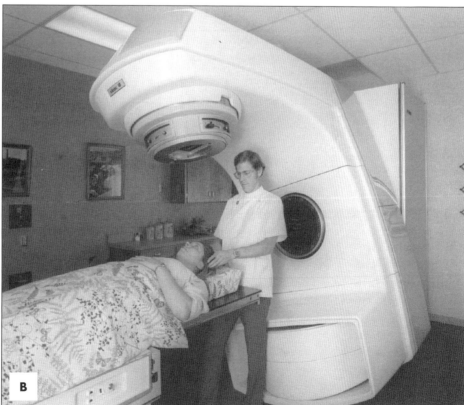

Figure 42. Linac with retractable beam stopper (Clinac 18). (a) Fully extended. (b) Retracted. (Courtesy of University of Arizona, Tucson, Arizona, and Varian Associates.)

13 FUNCTIONAL BLOCK DIAGRAM AND AUXILIARY SYSTEMS

A number of auxiliary systems are essential for operation, control, and monitoring of the linac treatment unit. These systems, together with the major components described earlier, are shown in Figure 43, a functional block diagram. The modulator cabinet (which may include the klystron) and control console, shown on the left, are located outside the treatment room; the stand, gantry, and treatment couch, shown on the right, are inside. The modulator cabinet may be placed inside the treatment room in some installations.

The modulator cabinet contains a pulsed power supply, as shown in Figure 43, which energizes the klystron and the electron gun when triggered by a timing pulse (Figure 44a) from the control console. The pulsed power supply provides a 120-kV pulse of approximately 5 microseconds duration to the klystron, which generates the microwave power, and a similar 18-kV pulse, which speeds electrons from the electron gun into the accelerator structure (Figures 44b, c, and d). The timing pulse rate, which is set by the radiation physicist, provides a convenient method of varying the linac output dose rate.

Electrons are injected into the structure on axis from the electron gun, as shown in the upper left of the gantry in Figures 15 and 43. The gun is pulsed with a negative 18-kV pulse. As a result, electrons enter the cavities with about 18 keV of energy and a velocity approaching one-fourth the speed of light.

The **vacuum system** provides the extremely low pressures needed for operation of the electron gun, accelerator structure, and bending magnet system. Without a vacuum, the electron gun would rapidly "burn out," like a lightbulb filament exposed to air. In addition, the accelerated electrons would collide with air molecules, deflecting them and reducing their energy, and the small, pencil-like beam of electrons would be broken up and diffused. The vacuum is maintained by an **electronic ion pump**. It was this latter development, more than any other, that transformed the linac from a laboratory instrument into a practical clinical tool. Earlier accelerator vacuum systems involved oil-based rotary and diffusion pumps that required significant maintenance.

The **pressure system** pressurizes the waveguide with sulfur hexafluoride, an insulating gas. This is needed to prevent electrical breakdown from the high-power microwave electric "E" fields.

A **cooling system**, providing temperature-controlled water, establishes the operating temperature of sensitive components and operates primarily to remove residual heat dissipated in other components.

Temperature control is particularly critical for the accelerator structure itself. Without it, the series of cavities composing the accelerator structure can change dimensions slightly. The effect of this is to "detune" them in the same way a musical instrument changes its pitch; they are then "off-frequency," and their acceleration capability is seriously impaired.

An **automatic frequency control (AFC)** system continuously senses the optimum operating frequency of the accelerator structure to maximize radiation output. It uses this information to "tune" the klystron or magnetron to this microwave frequency.

An elaborate **monitor** and **control system** maintains control of linac operation and patient treatment. It monitors operation to ensure proper linac performance and to ensure that the prescribed treatment is faithfully delivered in a safe manner. Deviations, depending on their nature and magnitude, will give rise to fault warning signals or termination of the treatment, when appropriate. The center of this monitor and control function is at the **control console** with connections to all other units. The control console provides status information on treatment modality, accessories in use, set dose, and dose delivered, interlock status, emergency off, as well as other data pertinent to linac operation and patient treatment. Frequently, the monitor function is directly linked to the control function and current status information is used in a feedback manner to maintain optimal performance.

A multitude of quantitative and procedural checks are incorporated in the console to ensure correct, safe operation. The digital logic circuits used in modern computers are the basis for these checking procedures. They can be carried out in a few seconds and are assessed automatically, prior to each treatment.

A counting system, tied to the dose monitor, terminates the treatment when the set dose **monitor units** (MU) are delivered. A backup timer is set to terminate treatment in the event of dose monitor system

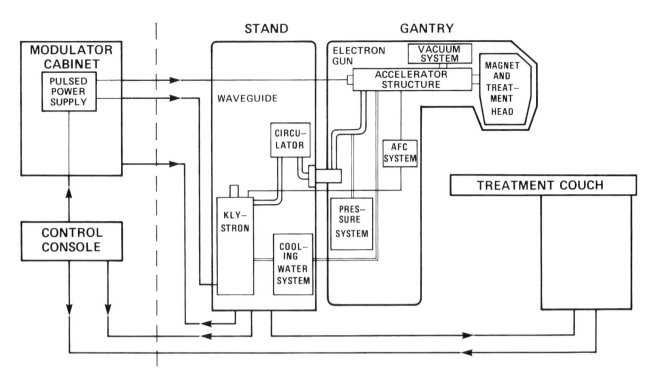

Figure 43. Block diagram of a high-energy bent-beam medical linac. Major components, auxiliary systems, and interconnections are identified. (Clinac 18, courtesy of Varian Associates.)

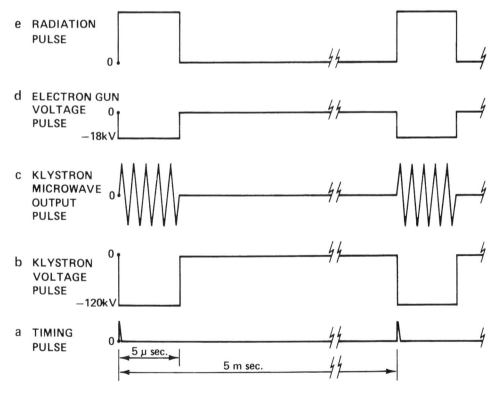

Figure 44. Timing diagram showing major time/event relationships for two sequential bursts of radiation. Note the two different abscissa time scales, one in microseconds (μ sec), the other milliseconds (msec).

failure. The radiation therapist monitors the treatment on a closed-circuit TV(see figure 4), and a two-way audio system permits communication with the patient.

14 OPERATIONAL REVIEW

An extensive treatment plan is prepared prior to treatment for each individual patient, usually consisting of a sequence of treatments carried out over a period of several weeks. The radiation oncologist specifies the tumor location and volume, treatment fields, and the prescribed dose for the patient's tumor. The number of monitor units to deliver one unit of absorbed dose in gray (Gy) varies for field size, modality, energy, and treatment distance. It is customary to define a dose monitor unit as the ionization produced under defined conditions. Generally, the linac is adjusted so that 1 (1.00) MU equals 1 (1.00) centigray at the depth of dose maximum (d_{max}) on the central axis in water for a 10 cm × 10 cm field at an SSD of 100 cm for a given energy and modality. For example, 150 MU would deliver 1.5 Gy for these conditions. The dose elsewhere is specified by isodose curves (actually surfaces), that is, curves of constant dose that are arbitrarily set to 100% where passing through d_{max}. For other conditions or locations, calibration factors must be applied to determine the required MU setting to deliver the prescribed dose.

The patient is positioned on the treatment couch by the radiation therapist; the gantry angle, collimator angle, field size, and treatment distance are set. Accessory beam-modifying devices such as blocks or wedges are attached and positioned. The radiation therapist then proceeds to the control console and sets the controls to deliver the daily dose for that treatment field. The radiation therapist must select the treatment modality: electrons or x-rays. The energy of the electrons and x-rays must also be selected.

In typical medical linacs, electrons used directly for treatment have energies from about 3 to 25 MeV. However, in the case of x-ray selection, a particular medical linac has only one or two energies, anywhere from 4 to 25 MV. For example, the Clinac 18 allows electron selection of 6, 9, 12, 15, or 18 MeV and produces x-ray beams of 6 or 10 MV.

Before the treatment begins, however, an internal check system is automatically activated that sequentially verifies linac operating parameters for correct values. In most units a method for testing the dosimetry system is used to ensure that the prescribed dose will be delivered. The treatment system may also include a computer-based **verify and record program**, which compares the treatment that has been set up with a record of the intended treatment (see Figure 4). This treatment assessment may include field size and collimator angulation, gantry angle, couch position, the daily dose for each field, and the preset monitor readings to provide that dose. Such verify and record programs identify setup errors prior to treatment so that they can be corrected when they exceed a preset magnitude, for example, more than one degree of arc. Such programs verify and record each treated field on a continuing, daily basis throughout the course of treatment. Medical accelerator safety considerations, procedures, and training are reviewed by Purdy et al. in the AAPM Task Group #35 Report.

Typically, the linac is pulsed several hundred times per second, with the exposure for each treatment field lasting a few minutes. When the "beam on" button is pushed, an elaborate sequence is initiated, in part described by the timing diagram of Figure 44. First, the modulator accumulates energy for the first pulse of radiation. It sends out two high-voltage pulses in unison: one to give the electrons leaving the electron gun their first boost of energy as they enter the accelerating structure (Figure 44d), the other to energize the klystron (Figure 44b). The klystron then delivers the microwave power to the accelerating structure (Figure 44c) and, in turn, to the electron beam emerging from the electron gun. Here, the intense "E" fields come into play, bunching the electrons and accelerating them to their final energy.

The electron beam next traverses the bending magnet and is directed on the x-ray target, or scattering foil, in the case of electron therapy. The emerging cone of radiation traverses the two monitor ionization chambers and is further shaped by the collimator and other beam-shaping devices.

Figure 44 summarizes pertinent time relationships for two sequential bursts of radiation. In this diagram of idealized timing, the linac is pulsed every 5 milliseconds, that is, 200 times per second. The timing pulse that initiates each sequence is very short, and all other pulses are of about 5 microseconds duration. During this 5 microsecond interval, 15,000 complete microwave cycles occur (3,000/µsec × 5). This microstructure is also present in the radiation burst (Figure 44e), but the timing details have been omitted for simplicity.

Verifying correct patient position with respect to the treatment beam can be carried out with "**port films,**" which are megavoltage x-ray radiographs showing the beam shape and size transmitted through the patient. The tumor visualization is poor, however, especially for high megavoltage energies. A number of auxiliary electronic portal imaging (EPI) systems have been developed to view real-time treatment port images (see figure 4).

The authors hope that this primer leads to a better understanding of medical linear accelerators, promotes clear communication between all personnel involved, and contributes to the safe and effective treatment of patients.

Bibliography

Karzmark, C.J. "Advances in Linear Accelerator Design for Radiotherapy." Medical Physics 11(2): 105–128, 1984.

Karzmark, C.J., C.S. Nunan, and E. Tanabe. Medical Electron Accelerators. McGraw-Hill, Inc., New York (1993).

Karzmark, C.J., and N.C. Pering. "Electron Linear Accelerators for Radiation Therapy: History, Principles and Contemporary Developments." Physics in Medicine and Biology 18:321–354 (1973).

Kramer, S., N. Suntharalingam, and G.F. Zinninger, Eds. High Energy Photons and Electrons: Proceedings of an International Symposium on the Clinical Usefulness of High-Energy Photons and Electrons (6-45 MeV) in Cancer Management. Thomas Jefferson University, Philadelphia, Pa., May 22–24, 1976. John Wiley and Sons, New York (1976).

Purdy, J.A. et al. Medical Accelerator Safety Considerations. Report of AAPM Radiation Therapy Committee, Task Group #35, Medical Physics 20(4): 1261–1275, 1993.

Tapley, Norah duV, Ed. Clinical Applications of the Electron Beam. John Wiley and Sons, New York (1976).

The Use of Electron Linear Accelerators in Medical Radiation Therapy: Physical Characteristics. HEW Publication (FDA) 768027 (1976).

Appendix

Representative Linac Treatment Units

Descriptions and photographs of a variety of medical linear accelerators have been solicited from various manufacturers. The manufacturer is solely responsible for the information provided. This section is provided for the education of the student and is not an endorsement of any product by the authors or publisher.

Elekta Oncology Systems Limited

Elekta Oncology Systems, formerly Philips Medical Systems, manufactures two types of linear accelerators, the SL75/5 and the SLi Series.

The SL75/5 is a single x-ray energy accelerator that may be set to 4.5 or 6 MV. The gantry is of the C-arm type with a 90° bending system. The control system is microprocessor based with an optional Verification and Recording system, Vericord S3, which provides a user interface that is compatible with the SLi interface.

The SLi Series is a range of computer-controlled accelerators. The series is based on a common design configuration and comprises the following models, all with triple x-ray energy and multiple electron energy choices. The configurations start at a basic entry level and, using a modular approach, customers can move forward to the full configuration as they require. See the configuration table below for a comparison between the two models.

The SLi accelerators are of drum gantry construction and use a traveling waveguide and the "Slalom" beam bending system, which results in a low isocentric height. The waveguide is guaranteed for 20 years, and the magnetron has a non-prorated warranty of 2 years. Rapid beam start-up characteristics and computer control combined with the integrated 80 leaf multileaf collimator, the MLCi, ensures that the SLi accelerators are well positioned for conformal techniques.

Configuration Table for SLi and SLi Plus

MACHINE	PHOTONS LOW	PHOTONS MEDIUM	PHOTONS HIGH	BASIC ELECTRONS	OPTIONAL ELECTRONS
SLi*	4,6	6,8	10,15	6,8,10	4,12,15,18
SLi plus**	4,6	6,8	10,15,18,25	6,8,10,12,15	4,18,20,22

*Standard is 1 photon from low or mid range and 3 electrons; option to increase to 2 or 3 photon energies (in the range of 4 to 15 MW); option to add electron energies up to a maximum of 7.

**Standard is 2 photons from any range and 5 electrons; option to add a 3rd photon energy (in the range of 4 to 25 MW); option to add electron energies up to a maximum of 9.

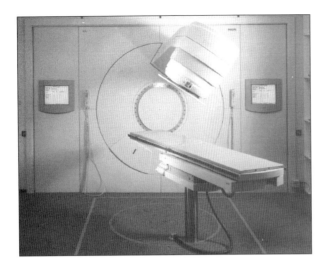

Elekta SL*i* Series

There are several design differences between Elekta Oncology Systems and other manufacturers' accelerators; the following are of particular interest.

Traveling Waveguide and Spectrometer Beam Bending System

All Elekta Oncology Systems accelerators have a high efficiency traveling waveguide structure for high dose rates and maximum versatility. The low vacuum requirement leads to easy replacement of vacuum components, e.g., demountable electron guns and rapid return to clinical use. The spectrometer type beam bending systems, 90° for the SL75/5 and the triple magnet 'Slalom' for the SL*i* Series, give very accurate control of energy.

100 cm Source Axis Distance— 124/118 cm Isocentric Height

Elekta Oncology Systems accelerators are designed for ease of use. The low isocentric height of 124 cm above floor level (118 cm for the SL75/5) enables operators to work at a convenient level without the need for step stools or elaborate floor modifications. This is a critical factor for accurate setup, ease of use, and improved throughput.

Automatic Wedge Filter System (SL75/5 and SL*i* Series)

The microprocessor control console or computer control systems used with the SL75/5 and SL*i* Series accelerators, respectively, enable operators to select the precise wedge angle required by automatically

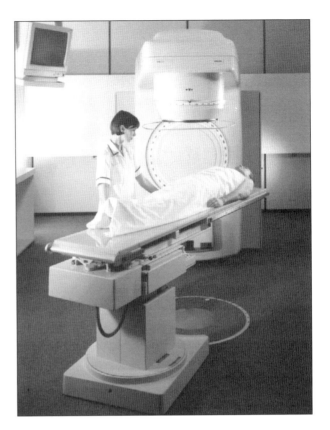

Elekta SL75/5

combining a 60° wedged field with an open field. By varying the dose with and without the wedge, any wedge angle between 0 and 60° can be obtained without the operator having to insert wedges manually.

SL*i* Series Control System

The control system of the SL*i* Series is computer based and forms an integral part of the linear accelerator, controlling both patient treatment and machine performance.

The patient prescription is stored within the system and includes not only gantry angle, field size, radiation modality, energy, and dose but also details of the planned series of treatments that can be customized to match the department practice.

Treatment is initiated when the operator selects the patient identifier, which triggers an automatic load of the prescription.

Slalom Beam Bending System

The slalom beam bending system consists of three in-line beam bending magnets. These are positioned along the evacuated flight tube through which the electron beam travels when it leaves the waveguide.

The first electromagnet acts as an energy analyzer and bends the electron beam through an angle of about 45°. It has shaped pole pieces to produce a dispersed beam having a spectrum of energies, with the high energy electrons on one side and the low on the other. The strength of the magnetic field is adjusted to transmit the desired mean energy electrons.

The second converging electromagnet reverses the 45° deflection and its pole pieces are shaped to start focusing the electron beam in two orthogonal directions as it enters the third magnet. The third magnet turns the electron beam through an angle of about 112° and also has shaped pole pieces that complete the two-dimensional focusing action started by the second magnet. The electrons are focused on a small area of the target, approximately 2 mm in diameter.

The Slalom system thus provides a very small diameter beam of electrons that is positionally fixed and inherently stable. The gun current is served by signals from the ion chamber to ensure constant energy control.

Networking for SLi Series and SL75/5

Several options are available to network the accelerator prescription databases and provide connectivity to external systems. These range from RTNet, which provides prescription data access between all connected accelerators, to transparent links to third party products that provide administrative functions.

Integrated Multileaf Collimator

The MLCi is an integrated device that fits into the same physical envelope as the standard radiation head. The prescription data for both the accelerator and the MLC are stored in the same file on the database so clinical operation is optimized.

The combined low isocentric height of the accelerator and the wide clearance between the end of the MLC and isocenter ensure optimum clinical use of the system.

> Elekta Oncology Systems Limited,
> Linac House, Fleming Way
> Crawley, West Sussex RH10 2RR
> United Kingdom

GE Medical Systems

With the acquisition of CGR in 1987, GE Medical Systems expanded to meet the needs of the radiation therapy community. The full range of products currently available has been developed from the years of experience gained by CGR-MeV in the field of particle accelerators. CGR-MeV pioneered their application of high energy x-ray and electron medical linear accelerators in the treatment of cancer with the Sagittaire™ in 1967. Sagittaires with 12 and 25 MV photons and 7 to 40 MeV electrons are still in clinical use today; however, they are no longer in production.

Today, GE continues to manufacture a full range of radiotherapy systems, including the Saturne 4™ series of accelerators.

High Energy Capabilities

The Saturne family of accelerators are useful for a broad range of clinical challenges. The optimized design of the standing-wave accelerating waveguide allows a unique range of energies for both magnetron and klystron powered systems. The streamlined magnetron powered Saturne 41 provides the user with two photon energies in a range from 4 MV to 15 MV, and 6 electron energies in a range from 4.5 MeV to 15 MeV. The Saturne 42, a medium range Klystron powered accelerator, provides users with 2 photon energies from 4 MV to 20 MV, and 8 electron energies in a range from 4.5 MeV to 21 MeV. The high energy, klystron powered Saturne 43 provides three photon energies from 4 MV to 25 MV, and 8 electron energies in a range from 4.5 MeV to 21 MeV.

Each Saturne system offers a number of capabilities designed to extend its clinical utility further; for example, clockwise and counterclockwise photon and electron arc therapy, multiple microprocessor-based control consoles, and automatic pre-setup are standard.

The Saturne series provides users with electron and photon beams meeting exceptionally high standards of homogeneity and purity.

The 270° achromatic beam path and the use of an automatically adjusted energy slit provide excellent electron energy definition. An optional high dose rate of 1,000 mu/min is available.

Other key features of the Saturne family include the following:

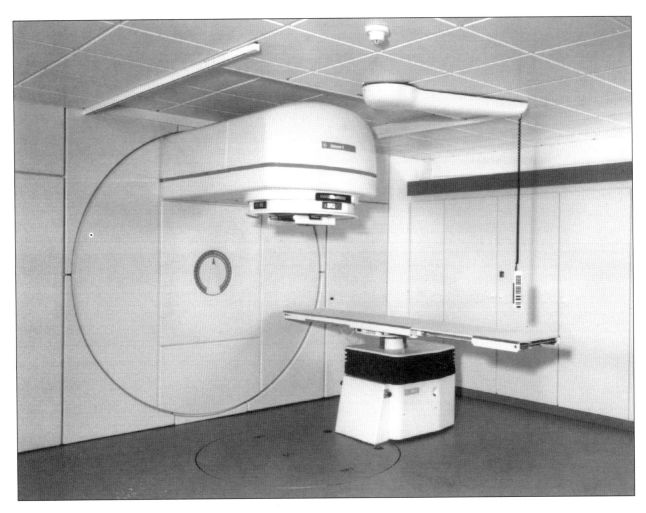

GE Saturne 4

Irradiation Fields Saturne series linacs offer exceptionally large, unclipped treatment fields; the fixed pre-collimator defines a 56 cm diameter circle at 1 m from the target.

In photon mode, the Saturne provides continuously variable, unclipped fields of up to 40×40 cm; in electron mode, up to 30×30 cm, via two pairs of mobile electron applicators.

Photon Mode A single, internal wedge filter, integrated with the beam limiting device, permits automatic wedge angles of up to $60°$ for maximum fields of 40×20 cm.

Electron Mode The double scattering foil design provides excellent homogeneity and minimum photon contamination over the available range of electron energies.

Dosimetry To promote symmetry, homogeneity, high and low dose output, and offset of accumulated dose rate, Saturne linacs employ two separate, open ion chambers associated with two independent dosimetry channels.

Beam Limiting Device (Collimator) The beam limiting system works in both symmetrical and asymmetrical modes. Asymmetry is possible on one pair of jaws up to the central axis of the field in both photon and electron modes.

Light Simulation To permit precise coincidence between the light field and the radiation field, a lamp housed in the target bar takes the place of the target during light simulation.

Commands in Treatment Room A newly designed control pendant allows all geometrical parameters to be set manually or automatically in the treatment room. A monitor within the treatment room provides the necessary display. All commands for the treatment table are also on the control pendant.

Control Console The control console has as its principal function the dialogue between operator and unit, as well as maintaining control of the following:

- geometric and dosimetric treatment parameters
- beam quality control
- safety device management

Two other computer functions are also included, one for physics, research, and testing and one for technical adjustments and maintenance. In addition, an InSite™ remote diagnostics connection is provided to allow maximized reliability through dedicated GE service engineers.

> GE Medical Systems
> W-412, P.O. Box 414
> Milwaukee, WI 53201-0414

Mitsubishi Electronics

Mitsubishi first developed medical linear accelerators in 1957. Throughout a 40-year period, Mitsubishi has produced high energy medical accelerators, research accelerators, and industrial accelerators. We developed our medical linear accelerators in cooperation with physicists and therapists and began production in 1965. There are now more than 400 clinical installations of Mitsubishi medical linear accelerators worldwide.

The Mitsubishi complete series of EXL computerized linear accelerators range from 4 MV photons to 18 MV photons with therapeutically useful electron energies up to 20 MeV. Continued research and development towards more clinically advanced technologies, such as precision multileaf collimation, stereotactic radiotherapy, real time portal imaging, and data management systems, has further supported and enhanced the EXL product line.

The EXL linear accelerators are designed to offer a wide range of photon energies from 4 MV to 18 MV

and electron energies from 4 MeV to 20 MeV. All EXL systems are composed of three major components (gantry, treatment table, and control console), and all utilize the well-proven standing wave accelerator guide technology, a demountable electron gun, and a 270° achromatic beam bending system. The EXL systems are available in both magnetron and klystron powered systems.

The optional asymmetric collimator facilitates complex field setup and offers extended travel beyond the central axis. Advanced technology allows the replacing of port film with real-time digital imaging. Patient positioning and treatment are displayed on-line at the digital portal imaging workstation, where image review, evaluation, and image post processing functions can be easily performed and stored.

The EXL computerized control console provides high photon dose rates up to 500 MU, high electron dose rates up to 1000 MU, advanced treatment modes, parameter display and adjustments, service and diagnostics capability, remote gantry controls at computer console, automatic or manual parameter input, and an included basic verification system.

> Mitsubishi Electronics America, Inc.
> Radiation Oncology Group, Ste. 100
> 2 Tower Bridge, 1 Fayette St.
> Conshohocken, PA 19428

Siemens Medical Systems-OCS

Siemens Medical Systems, Inc., Oncology Care Systems Group is located in Concord, California. Siemens manufactures a family of Mevatron linear accelerators that was recently enhanced by its latest edition, the Primus linear accelerator. Throughout the product line, dual photon and electron capabilities are available.

Due to the unified design, upgrades are possible within the product line. All accelerators are digitally controlled, allowing upgrades like virtual wedges or remote diagnostics. The control architecture is open in design to allow for implementing upgrades as technology evolves.

The Primus linear accelerator is configured for conformal therapy. Virtual Wedge, 3D Multileaf Collimator, Asymmetric Collimators, the Simtec (Sequential Intensity Modulation Technology) module, and total compatibility with the Lantis Data Management System are options to the Primus accelerator.

The Primus linear accelerator features the industry's most compact design for a high energy accelerator. This was made possible due to a quantum leap in technology by using the latest solid-state technology throughout the modulator assembly. This technology allows for an increase in dose rate, reduces the space requirements, and shortens the installation time.

Revolutionary solid-state technology, advanced design internal cooling, and insulated covers all contribute to the extremely quiet operation of the Primus accelerator. The cover design provides for maximum clearance for optimal couch rotation for non-coplanar beams, which is important due to the complexity of 3D treatment plans.

Most digital linear accelerators can be upgraded with virtual wedge, allowing the generation of wedged dose profiles as a result of jaw motions during radiation. Every virtual wedge is verified with the dynamic icon displaying the values for monitor units versus jaw position. This provides a real-time verification ensuring every wedge angle is delivered correctly. In order to ease the treatment planning for virtual wedge treatments, the wedge correction factor does not vary with field size.

Siemens offers a 3D multileaf collimator. The multileaf collimator is double focused, which means it is divergent in X and Y directions. The double-focused Siemens multileaf collimator is fully integrated into the control system, allowing for automated setup via the Lantis Data Management System. All treatment parameters, including the multileaf collimator positions, are verified and documented. The multileaf collimator covers a field size of 40 cm × 40 cm with a resolution of 1 cm for the inner 27 leaves and 6.5 cm for the outer leaves. Each leaf has a travel range of 30 cm independent of the position of the adjacent leaf.

The multileaf collimator was also integrated into the defining head and replaces one set of jaws. This allows for optimal clearance between the accessory holder and the isocenter. Clearance around the patient is an important clinical factor when you

Siemens Multileaf Collimator

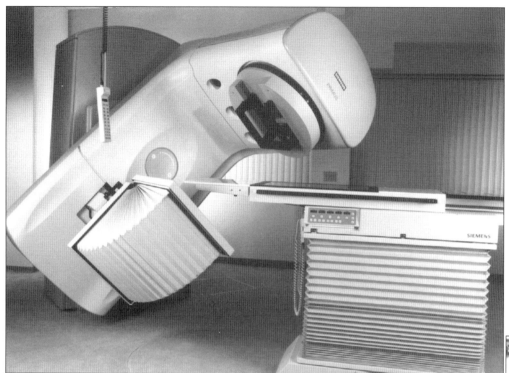

Siemens PRIMUS Linear Accelerator with
ZXT Treatment Table and BEAMVIEW
PLUS Real-time Portal Imaging System

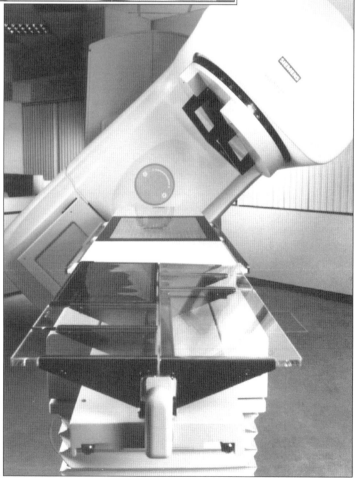

consider techniques for conformal therapy, including non-coplanar beams and the automatic delivery of a sequence of fields.

Design criteria for the multileaf collimator include: optimal clearance and low leakage values combined with high long-term reliability. The setup of leaf position can be achieved with the Beamshaper, a link to a treatment planning system and in-room hand control.

A ZXT patient treatment table provides extended vertical motions, and a number of modern accessories are available. Also, every accelerator can be equipped with the BEAMVIEW PLUS electronic portal imaging system.

All accelerators have a verification system interface. Siemens provides the LANTIS Data Management System to network your radiation oncology department fully. Verification, auto-setup of the accelerator, scheduling, generation of reports and transcriptions, interfaces to simulators and treatment planning systems, as well as image management are some of the LANTIS functions. This complete solution for radiation oncology optimizes the delivery of conformal therapy for every patient.

Siemens Medical Systems-OCS
4040 Nelson Ave.
Concord, CA 94520

Varian Oncology Systems

Varian Oncology Systems produces a wide array of linear accelerators and accessories for oncology departments.

600C

The Clinac 600C radiotherapy accelerator generates 4 and 6 MV x-ray beams. It provides rectangular symmetric and asymmetric fields and sophisticated radiotherapy treatments, including multiport wedged and nonwedged treatments and rotational treatments.

The 600C can treat fields ranging in size from 0.5×0.5 cm to 40×40 cm at a 100 cm target-to-skin distance. The dose rate for stationary therapy is variable in five equal increments from a minimum rate of 50 monitor units per minute up to 250 MU/minute, depending on the system configuration. X-ray arc

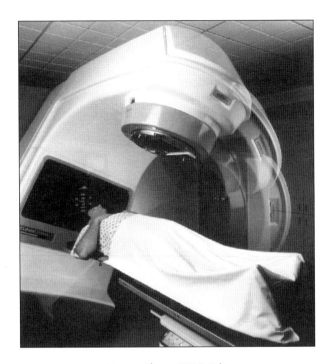

Varian Clinac 2300 C/D

therapy can be performed with dose rates ranging from 0.5 to 10 MU/degree increments. A 600C/D is also available that provides the capability of performing dynamic beam and complex treatments, in addition to conventional techniques. Its dose rate ranges from 100 to 600 MU/minute.

2100C

The Clinac 2100C offers a diverse array of energy options for treating patients, regardless of size and tumor location. There is a choice of six photon combinations between 4 and 23 MV, each with two widely separated x-ray energies, made possible by Varian's unique energy switch. Five electron energies between 4 and 20 MeV are also available. A 2100C/D is also available that provides the capability for performing dynamic beam and complex treatments, in addition to conventional techniques. Its dose rates range from 100 to 1000 MU/minute.

2300C/D

The Clinac 2300C/D offers conventional and conformal treatments, backed by a full range of treatment beams and dose rates from 100 to 1000 Monitor Units per minute. A selection of dual photon energies between 6 and 25 MV and six electron energies between 4 and 22 MeV is available.

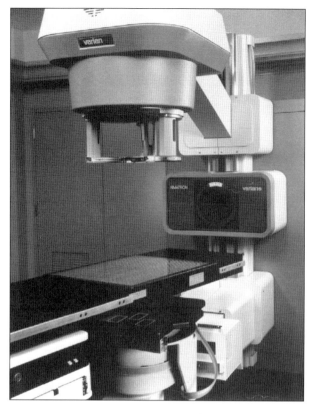

Varian Ximatron Simulator

Varian Clinac 2300C/D with MLC

600SR

The Clinac 600SR is a dedicated system integrating beam delivery, stereotactic immobilization instrumentation (developed by Radionics, Inc.), and the XKnife treatment planning system pioneered at Brigham and Women's Hospital and Joint Center for Radiation Therapy, Harvard Medical School.

Varian Multileaf Collimator

Varian's MLC replaces many custom shielding blocks and allows fast and easy setup for multiple static ports. This fits on the Clinac 600C, 2100C, and 2300C/D.

Varian PortalVision

PortalVision (PV) creates high-quality images that are easily acquired and matched to your simulator films. The imaging package includes a remote viewing station.

Other Related Products

Ximatron Simulator, Ximatron/CT Option Varian produces Simulators and CT Scanning Options for ensuring high quality treatments. The Ximatron CX offers a high frequency fluoroscopic system and a 12 inch image intensifier with a microprocessor control system. The Ximatron/CT Option provides therapeutic-quality CT images at simulation for use in treatment planning.

VariSource VariSource is Varian's high dose rate remote afterloader for brachytherapy. Operating with needles as small as 0.8 mm diameter and catheters as small as 1.6 diameter, patient discomfort is greatly reduced and treatment tolerance is significantly improved. VariSource's high speed drive wheel and tension belt system offers precise source positioning and fast source movement (up to 50 cm per second) minimizing patient transit dose.

CADPLAN Treatment Modeling Workstation CADPLAN Treatment Modeling Workstation offers a wide range of tools for image processing to help in the definition of the target volume and other internal structures. For treatment modeling, users have 3D visualization and 3D dose display options, which help to predict the optimum treatment plan.

Varis Varis is a comprehensive radiotherapy information management system. Modules are available for Scheduling, Simulation, Treatment, Images, Reports, Charges, OncoLog (provides tumor registry, outcomes analysis, and reporting capabilities), Med-Oncology (manages medical oncology information including clinical data, notes, and charts), and Link (which provides industry-standard connectivity to external information systems such as hospital billing or scheduling systems).

Varian Oncology Systems
3045 Hanover St.
Palo Alto, CA 94304

Index